Organic Structures from 2D NMR Spectra

Organic Structures
from 2D NMR Spectra

L. D. Field, H. L. Li and A. M. Magill
School of Chemistry, University of New South Wales, Australia

WILEY

This edition first published 2015
© 2015 John Wiley & Sons Ltd

Registered office
John Wiley & Sons Ltd, The Atrium, Southern Gate, Chichester, West Sussex, PO19 8SQ, United
Kingdom

For details of our global editorial offices, for customer services and for information about how to
apply for permission to reuse the copyright material in this book please see our website at
www.wiley.com.

Library of Congress Cataloging-in-Publication Data

Field, L. D.
  Organic structures from 2D NMR spectra / L.D. Field, H.L. Li, and A.M. Magill.
    pages cm
  Includes bibliographical references and index.
  ISBN 978-1-118-86894-2 (pbk.)
 1. Nuclear magnetic resonance spectroscopy.   2. Spectrum analysis.   I. Li, H. L. (Hsiu L.)
II. Magill, A. M. (Alison M.)   III. Title.   IV. Title: 2D NMR spectra.
  QD96.N8F54 2015
  543′.66–dc23
                                                                      2015008088

A catalogue record for this book is available from the British Library.

ISBN: 9781118868942

1   2015

CONTENTS

Contents

PREFACE

Obtaining structural information from spectroscopic data is an integral part of organic chemistry courses at all universities. At this time, NMR spectroscopy is arguably the most powerful of the spectroscopic techniques for elucidating the structure of unknown organic compounds, and the method continues to evolve over time.

This text *Organic Structures from 2D NMR Spectra* builds on the popular series *Organic Structures from Spectra*, which is now in its fifth edition. The aim of *Organic Structures from Spectra* is to teach students to solve simple structural problems efficiently by using combinations of the major spectroscopic and analytical techniques (UV, IR, NMR and mass spectroscopy). Probably the most significant advances in recent years have been in the routine availability of quite advanced 2D NMR techniques. This text deals specifically with the use of more advanced 2D NMR techniques, which have now become routine and almost automatic in almost all NMR laboratories.

In this book, we continue the basic philosophy that learning how to identify organic structures from spectroscopic data is best done by working through examples. Solving real problems as puzzles is also addictive – there is a real sense of achievement, understanding and satisfaction. About 70% of the book is dedicated to a series of more than 60 graded examples ranging from very elementary problems (designed to demonstrate useful problem-solving techniques) through to very challenging problems at the end of the collection.

The underlying theory has been kept to a minimum, and the theory contained in this book is only sufficient to gain a basic understanding of the techniques actually used in solving the problems. We refer readers to other sources for a more detailed description of both the theory of NMR spectroscopy and the principles underpinning the NMR experiments now in common use.

The following books are useful sources for additional detail on the theory and practice of NMR spectroscopy:

(i) T. D. W. Claridge, *High-Resolution NMR Techniques in Organic Chemistry*, 2nd edition, Elsevier, Amsterdam, 2009. ISBN 978-0-08-054628-5.

(ii)   J. Keeler, *Understanding NMR Spectroscopy*, 2nd edition, John Wiley & Sons, UK, 2010. ISBN 978-0-470-74609-7.

(iii)   H. Friebolin, *Basic One- and Two-Dimensional NMR Spectroscopy*, 5th edition, Wiley-VCH, Weinheim, 2011. ISBN 978-3-527-32782-9.

(iv)   H. Günther, *NMR Spectroscopy: Basic Principles, Concepts and Applications in Chemistry*, 3rd edition, Wiley-VCH, Weinheim, 2013. ISBN 978-3-527-33000-3.

In this book, the need to learn data has been kept to a minimum. It is more important to become conversant with the important spectroscopic techniques and the general characteristics of different types of organic compounds than to have an encyclopaedic knowledge of more extensive sets of data. The text does contain sufficient data to solve the problems, and again there are other excellent sources of data for NMR spectroscopy.

The following collections are useful sources of spectroscopic data on organic compounds:

(i)   http://riodb01.ibase.aist.go.jp/sdbs/cgi-bin/cre_index.cgi?lang=eng, maintained by the National Institute of Advanced Industrial Science and Technology, Tsukuba, Ibaraki, Japan.

(ii)   http://webbook.nist.gov/chemistry/, which is the NIST Chemistry WebBook, NIST Standard Reference Database Number 69, June 2005, Eds. P. J. Linstrom and W. G. Mallard.

(iii)   E. Pretch, P. Bühlmann and M. Badertscher, *Structure Determination of Organic Compounds, Tables of Spectral Data*, Springer-Verlag, Berlin/Heidelberg, 2009. ISBN 978-3-540-93810-1.

ASSUMED KNOWLEDGE

The book assumes that students have completed an elementary organic chemistry course, so there is a basic understanding of structural organic chemistry, functional groups, aromatic and non-aromatic compounds, stereochemistry, etc. It is also assumed that students already have a working knowledge of how various spectroscopic techniques (UV, IR, NMR and mass spectroscopy) are used to elucidate the structures of organic compounds.

The following books are useful texts dealing with the elucidation of the structures of organic compounds by spectroscopy:

(i) L. D. Field, S. Sternhell and J. R. Kalman, *Organic Structures from Spectra*, 5th edition, John Wiley & Sons, UK, 2013. ISBN 978-1-118-32545-2.

(ii) R. M. Silverstein, F. X. Webster, D. J. Kiemle and D. L. Bryce, *Spectrometric Identification of Organic Compounds*, 8th edition, John Wiley & Sons, USA, 2014. ISBN 978-0-470-61637-6.

STRUCTURE OF THE BOOK

- *Chapter 1* deals with the basic physics of the NMR experiment and the hardware required to acquire NMR spectra.

- *Chapter 2* deals with the general characteristics of NMR spectroscopy for commonly observed nuclei. While most NMR deals with ^{1}H or ^{13}C NMR spectroscopy, this chapter also provides an introduction to ^{19}F, ^{31}P and ^{15}N NMR.

- *Chapter 3* deals with 2D NMR spectroscopy. First the principles, and then a basic description of the commonly used 2D NMR experiments – COSY, NOESY, TOCSY, INADEQUATE, HSQC/HMQC and HMBC.

- *Chapter 4* covers a group of special topics which are important in interpreting NMR spectra. Topics include (i) the common solvents used for NMR; (ii) the standard reference materials used for the observation of the spectra of different nuclei; (iii) the effects of molecular exchange and molecular motion on NMR spectra; and (iv) the effect of chirality on NMR spectra.

- *Chapter 5* contains two worked solutions as an illustration of a logical approach to solving problems. However, with the exception that we insist that students should perform all routine measurements first, we do not recommend a mechanical attitude to problem solving – intuition has an important place in solving structures from spectra.

INSTRUMENTATION

The NMR spectra presented in the problems contained in this book were obtained under conditions stated on the individual problem sheets. Spectra were obtained on the following instruments:

(i)   300 MHz ^{1}H NMR spectra, 75 MHz ^{13}C NMR spectra and 283 MHz ^{19}F spectra on a Bruker DPX-300 spectrometer;

(ii)   400 MHz ^{1}H NMR spectra, 100 MHz ^{13}C NMR spectra and 376 MHz ^{19}F spectra on Bruker Avance III 400 spectrometers;

(iii)   500 MHz ^{1}H NMR and 125 MHz ^{13}C NMR spectra on a Bruker Avance III 500 spectrometer;

(iv)   600 MHz ^{1}H NMR and 150 MHz ^{13}C NMR spectra were obtained on Avance III 600 or Avance III HD 600 Cryoprobe spectrometers.

There is a companion Instructor's Guide which provides a comprehensive step-by-step solution to every problem in the book.

Bona fide instructors may obtain a list of solutions (at no charge) by emailing the authors at L.Field@unsw.edu.au or fax (+61 2 9385 8008).

We wish to thank Dr Donald Thomas and Dr James Hook at the Mark Wainwright Analytical Centre at the University of New South Wales, and Dr Joanna Cosgriff and Dr Roger Mulder at CSIRO Materials Science and Engineering who helped to assemble the additional samples and spectra used in this book. Thanks are also due to Dr Samantha Furfari and Dr Manohari Abeysinghe who helped with the synthesis of several of the compounds used in the problems.

L. D. Field
H. L. Li
A. M. Magill
January 2015

LIST OF FIGURES

LIST OF TABLES

1 NMR Spectroscopy Basics

1.1 THE PHYSICS OF NUCLEAR SPINS

Any nucleus that has an odd number of protons and/or neutrons has a property called "nuclear spin". Such nuclei are termed "NMR-active nuclei" and, in principle, these nuclei can be observed by Nuclear Magnetic Resonance (NMR) spectroscopy.

Any nucleus that has an even number of protons *and* an even number of neutrons has no nuclear spin and cannot be observed by NMR. Nuclei with no nuclear spin are "NMR-silent nuclei". Common nuclei that fall into the NMR-silent category include carbon-12 and oxygen-16. Fortunately, with a few exceptions, most elements do have at least one isotope that has a nuclear spin, and so while ^{12}C and ^{16}O are NMR-silent, we can observe NMR spectra for the less abundant isotopes of carbon and oxygen, ^{13}C and ^{17}O. So even the elements where the most abundant isotope is NMR-silent can usually be observed via one or more of the less abundant isotopes.

Each nucleus has a unique nuclear spin, which is described by the spin quantum number, I. Nuclear spin is quantised, and I has values of 0, $\frac{1}{2}$, 1, $\frac{3}{2}$ *etc*. NMR-silent nuclei have $I = 0$. Each nuclear spin also has a magnetic moment, μ. The nuclear spin and the magnetic moment are related by Equation 1-1:

$$\mu = \gamma I \qquad\qquad (1\text{-}1)$$

The constant of proportionality, γ, is known as the *magnetogyric ratio*, and γ is unique for each NMR-active isotope. Table 1-1 provides a summary of the nuclear spins of some of the common NMR-active nuclei.

The combination of spin and charge means that NMR-active nuclei behave like small magnets and when a nucleus with a nuclear spin I is placed in an external magnetic field, that nucleus may assume one of $2I + 1$ orientations relative to the direction of the applied field.

Organic Structures from 2D NMR Spectra. L. D. Field, H. L. Li and A. M. Magill
© *2015 John Wiley & Sons, Ltd. Published 2015 by John Wiley & Sons, Ltd.*

Table 1-1 **Nuclear spins and magnetogyric ratios for some common NMR-active nuclei.**

Nucleus	Spin I	Natural Abundance (%)	Magnetogyric Ratio ($\gamma \times 10^7$ rad/T/s)
^{1}H	$^1/_2$	99.98	26.75
^{2}H	1	0.015	4.11
^{10}B	3	19.58	2.87
^{11}B	$^3/_2$	80.42	8.58
^{13}C	$^1/_2$	1.108	6.73
^{14}N	1	99.63	1.93
^{15}N	$^1/_2$	0.37	−2.71
^{17}O	$^5/_2$	0.037	−3.63
^{19}F	$^1/_2$	100.0	25.17
^{29}Si	$^1/_2$	4.7	−5.31
^{31}P	$^1/_2$	100.0	10.83

So, for a nucleus with $I = {}^1/_2$ like ^{1}H or ^{13}C, there are two possible orientations, which can be pictured as having the nuclear magnet aligned either parallel or antiparallel to the applied field. For nuclei with $I = 1$ there are three possible orientations; for nuclei with $I = {}^3/_2$ there are four possible orientations and so on.

$$I = {}^1/_2 \qquad I = 1 \qquad I = {}^3/_2$$

The various orientations of a nuclear magnet in a magnetic field are of unequal energy, and the energy gap (ΔE) is proportional to the strength of the applied magnetic field ($\boldsymbol{B}_0$) according to Equation (1-2:

$$\Delta E = \frac{h\gamma \boldsymbol{B}_0}{2\pi} \tag{1-2}$$

where h is the Planck constant.

Nuclei in a lower energy orientation can be excited to the higher energy orientation by a radiofrequency (Rf) pulse of the correct frequency (v) according to Equation (1-3:

$$v = \frac{\Delta E}{h}$$ (1-3)

It follows from Equations (1-2 and (1-3 that the fundamental equation that relates frequency (v) to magnetic field strength (B_0) is Equation (**1-4** which is known as **the Larmor Equation**:

$$v = \frac{\gamma B_0}{2\pi}$$ (1-4)

The Larmor equation specifies that the frequency required to excite an NMR-active nucleus is proportional to the strength of the magnetic field and to the magnetogyric ratio of the nucleus being observed. For magnetic fields that are currently accessible routinely for NMR spectroscopy (up to about 21 T), the frequencies required to observe most common NMR-active nuclei fall in the Rf range of the electromagnetic spectrum (up to about 900 MHz).

Table 1-2 summarises the NMR frequencies of common NMR-active nuclei.

Table 1-2   Resonance frequencies for some common NMR-active nuclei in different magnetic fields.

Nucleus	NMR Frequency (MHz) at 4.698 T	NMR Frequency (MHz) at 9.395 T	NMR Frequency (MHz) at 18.79 T
^{1}H	200.0	400.0	800.0
^{2}H	30.7	61.4	122.8
^{10}B	21.5	43.0	86.0
^{11}B	64.2	128.3	256.6
^{13}C	50.3	100.6	201.1
^{14}N	14.4	28.9	57.8
^{15}N	20.3	40.5	81.0
^{17}O	27.1	54.2	108.5
^{19}F	188.2	376.3	752.6
^{29}Si	39.7	79.4	158.9
^{31}P	81.0	161.9	323.8

1.2 BASIC NMR INSTRUMENTATION AND THE NMR EXPERIMENT

Samples for NMR spectroscopy are typically liquids (solutions) or solids. In order to observe Nuclear Magnetic Resonance, the sample must be placed in a strong magnetic field.

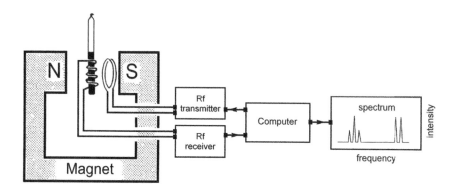

Magnets for NMR spectroscopy may be either permanent magnets or electromagnets. Most modern magnets are electromagnets based on superconducting solenoids, cooled to liquid helium temperature.

NMR spectrometers require an Rf transmitter which can be tuned to the appropriate frequency for the nucleus one wishes to detect (Equation (**1-4**) and an Rf detector or receiver to observe the Rf radiation absorbed and emitted by the sample. In most modern instruments, the Rf transmitter and the Rf receiver are controlled by a computer and the detected signal is captured in a computer which then allows processing and presentation of the data for analysis.

2 One-Dimensional Pulsed Fourier Transform NMR Spectroscopy

A short pulse of radiofrequency radiation will simultaneously excite all of the nuclei whose resonance frequencies are close to the frequency of the pulse. If a sample placed in a magnetic field of 9.395 T contains ^{31}P nuclei, then a pulse whose frequency is close to 161.9 MHz will excite all of the ^{31}P nuclei in the sample. Typically, the excitation pulse is very short in duration (microseconds). Once the pulse is switched off, the magnetisation which builds up in the sample begins to decay exponentially with time. A pulsed NMR spectrometer measures the decrease in sample magnetisation as a function of time, and records the *free-induction decay* (FID) (Figure 2-1).

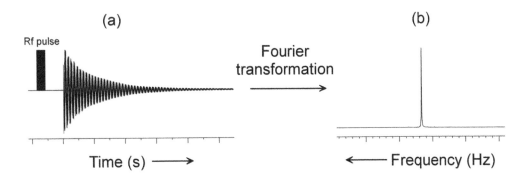

Figure 2-1 ^{1}H NMR spectra: *(a) time domain spectrum (FID); (b) frequency domain spectrum obtained after Fourier transformation of (a).*

The FID is a time domain signal (*i.e.* a signal whose amplitude is a function of time), and contains information for each resonance in the sample, superimposed on the information for all the other resonances. The FID signal may be transformed into the more easily interpreted frequency domain spectrum (*i.e.* a signal whose amplitude is a function of frequency), by a mathematical procedure known as ***Fourier transformation*** (FT). The frequency domain spectrum is the typical NMR spectrum that is used to provide information about chemical compounds. An NMR spectrum which contains intensity information as a function of one frequency domain is termed a *one-dimensional (1D) NMR spectrum.*

Organic Structures from 2D NMR Spectra. L. D. Field, H. L. Li and A. M. Magill
© 2015 John Wiley & Sons, Ltd. Published 2015 by John Wiley & Sons, Ltd.

There are typically multiple signals in any sample and the FID is then a complex superposition of all signals from the sample. The FT then provides a frequency domain spectrum with multiple resonances. The magnetisation in the sample decays back to equilibrium, typically over a period of seconds, by processes generally known as **relaxation**. The NMR experiment only works because there are mechanisms that restore the system back to equilibrium once it has been excited by absorption of Rf energy.

After a suitable delay to let the sample relax, the excitation pulse is repeated and another FID recorded. The FIDs collected can be added together to improve the intensity of the signal in the final spectrum.

For organic liquids and samples in solution, it may take several seconds for the system to relax. In the presence of paramagnetic impurities or in very viscous solvents, relaxation can be very efficient and, as a consequence, NMR spectra obtained become broadened.

If relaxation is too efficient (*i.e.* it takes a very short time for the nuclear spins to relax after being excited in an NMR experiment), the lines observed in the NMR spectrum are very broad. If relaxation is too slow (*i.e.* it takes a long time for the nuclear spins to relax after being excited in an NMR experiment), the resonances are sharp but then there must be a longer delay between pulses.

Not all NMR-active nuclei are easily observed using NMR spectroscopy:

i. Some nuclei suffer from a very low natural abundance, which simply means the concentration of NMR-active nuclei in a sample is low and the signal is weak.

ii. Nuclei with $I > \frac{1}{2}$ have an electric quadrupole which broadens NMR signals and makes spectra more difficult to observe. In contrast, those nuclei with $I = \frac{1}{2}$ typically give rise to signals which are sharp and easily observed. ^{1}H, ^{13}C, ^{19}F and ^{31}P all have $I = \frac{1}{2}$ and are the most commonly observed nuclei by NMR spectroscopy.

iii. Equation (1-2 indicates that ΔE is proportional to the strength of both the magnetic field and the magnetogyric ratio of the nucleus being observed. The intensity of the NMR signal depends on the population difference between the states – larger ΔE means a larger population difference and a stronger observed NMR signal. Nuclei with a low magnetogyric ratio give rise to only a small ΔE, which results in poor sensitivity.

iv.   Nuclei which are associated with a paramagnetic atom, *i.e.* where there are unpaired electrons, relax very efficiently and give rise to NMR signals which are broadened and more difficult to observe.

2.1 THE CHEMICAL SHIFT

While the Larmor equation and the information in Table 1-2 provide the broad distinction between the isotopes of different elements, the chemical significance of NMR spectroscopy relies on the subtle differences between nuclei of the same isotope which are in chemically different environments.

All ^{1}H nuclei in a sample are not necessarily equivalent, and the chemical environment that each ^{1}H finds itself in within the structure of the molecule determines its exact resonance frequency. Each nucleus is screened or shielded from the applied magnetic field by the electrons that surround it. Unless two ^{1}H environments are precisely identical (by symmetry) *their resonance frequencies must be slightly different.* Nuclei that are close to strongly electronegative functional groups have the local electronic environment distorted and may have less electron density to screen or shield them from the magnetic field and the nuclei are said to be **deshielded**. Nuclei that are in electron-rich sections of a molecule have more electron density to screen or shield them from the magnetic field and the nuclei are said to be **shielded**.

A typical NMR spectrum is a graph of resonance frequency against intensity. The frequency axis is calibrated in dimensionless units called "parts per million" (abbreviated to ppm). The chemical shift scale in ppm, termed the δ scale, is usually calibrated relative to the signal of a reference compound whose frequency is set at 0 ppm. For ^{1}H NMR spectroscopy, the reference is the proton resonance of tetramethylsilane ($Si(CH_3)_4$, TMS) and for ^{13}C NMR spectroscopy the reference is the carbon resonance of TMS. The frequency difference between the resonance of a nucleus and the resonance of the reference compound is termed the **chemical shift** (Equation 2-1).

$$\text{Chemical shift } (\delta) \text{ in ppm} \quad = \quad \frac{\text{Frequency difference from TMS in Hz}}{\text{Spectrometer frequency in MHz}} \qquad (2\text{-}1)$$

Note that for a spectrometer operating at 500 MHz, 1 ppm corresponds to 500 Hz, *i.e.* for a spectrometer operating at *x* MHz, 1 ppm always corresponds to exactly *x* Hz.

For the majority of organic compounds, the chemical shift range for ¹H covers approximately 0–10 ppm (from TMS) and the chemical shift range for ¹³C covers approximately 0–220 ppm (from TMS). By convention, the δ scale runs (with increasing values) from right to left with the signals of the most shielded nuclei at the right hand end of the spectrum and the least shielded nuclei to the left.

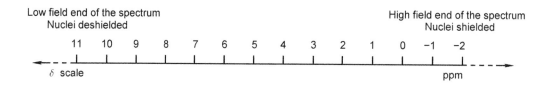

Any effect which alters the density or spatial distribution of electrons around a ¹H nucleus will alter the degree of shielding and hence its chemical shift. ¹H chemical shifts are sensitive to both the hybridisation of the atom to which the ¹H nucleus is attached (sp², sp³, *etc.*) and to electronic effects (the presence of neighbouring electronegative/electropositive groups).

The chemical shift of a nucleus reflects its local environment in a molecular structure and this makes NMR spectroscopy a powerful tool for obtaining structural information.

π-Electrons in organic structures generate a local magnetic field which can shield or deshield nearby NMR-active nuclei. Functional groups such as double and triple bonds, aromatic rings and carbonyl, nitro and nitrile groups distort the local magnetic field and these types of functional groups are termed *magnetically anisotropic* groups.

In aromatic rings, for example, the circulation of the π-electrons induces a small, localised magnetic field which deshields any nuclei which are in the plane of the aromatic ring and shields nuclei which are in the zone above or below the plane of the aromatic ring.

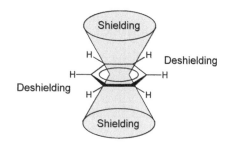

Shielding and deshielding by an aromatic ring is clearly illustrated by the compound on the right, in which the aromatic protons are deshielded, and resonate at 6.86 ppm, while the central bridge-head proton experiences significant shielding and resonates at −4.03 ppm, well upfield of most other proton signals.*

Alkenes and carbonyl compounds also display deshielding effects for protons directly bound to the functional group, while the terminal protons of alkynes fall in the shielding zone of the triple bond.

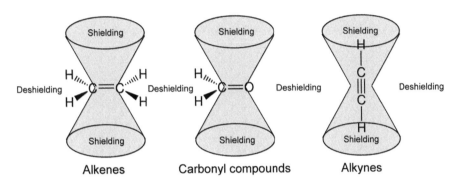

| Alkenes | Carbonyl compounds | Alkynes |

2.2 ^{1}H NMR SPECTROSCOPY

Proton NMR is the most commonly acquired type of NMR spectrum. Almost all organic compounds contain hydrogen; ^{1}H is the overwhelmingly most abundant isotope of hydrogen and ^{1}H is amongst the most sensitive nuclei to observe by NMR spectroscopy.

2.2.1 Chemical Shifts in ^{1}H NMR Spectroscopy

The chemical shifts for protons in organic compounds are summarised in Figure 2-2. A significant amount of information about the functional groups contained in a molecule can be deduced simply from the chemical shift ranges of the protons it contains.

---

* Pascal, R. A., Jr.; Winans, C. G.; Van Engen, D. Small, strained cyclophanes with methine hydrogens projected toward the centers of aromatic ring. *J. Am. Chem. Soc.*, **1989**, *111*, 3007–3010.

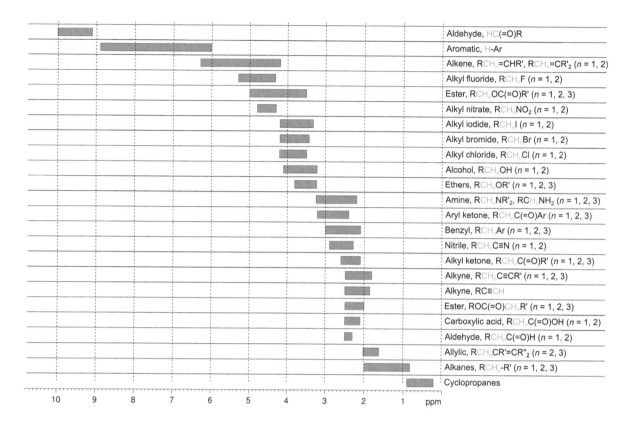

Figure 2-2   Approximate ^{1}H chemical shift ranges for protons in organic compounds.

2.2.2   Spin–Spin Coupling in ^{1}H NMR Spectroscopy

Most organic molecules contain more than one magnetic nucleus (*e.g.* more than one ^{1}H). When one NMR-active nucleus can sense the presence of other NMR-active nuclei ***through the bonds of the molecule*** the signals will exhibit fine structure (*splitting or multiplicity*).

Signal multiplicity arises because the energy of a nucleus, which can sense the presence of other magnetic nuclei, is perturbed slightly by the spin states of those nuclei.

The presence (or absence) of splitting due to spin–spin coupling provides valuable structural information when correctly interpreted. **Spin–spin coupling** is transmitted through the bonds of a molecule and so it is not observed between nuclei in different molecules. The effect of coupling falls off quite rapidly as the number of bonds between the coupled nuclei increases.

Signal multiplicity – the n+1 rule*.* Spin–spin coupling gives rise to multiplet splittings in ^{1}H NMR spectra. The NMR signal of a nucleus coupled to ***n*** equivalent hydrogens will be split into a multiplet with (***n***+1) lines (Figure 2-3). The CH$_2$ signal of bromoethane (H$_1$) is split into a multiplet with four lines by coupling with the three protons on the adjacent carbon ($n+1 = 4$). The CH_3 signal (H_2) is split into a multiplet with three lines by coupling with the two protons on the adjacent carbon ($n+1 = 3$).

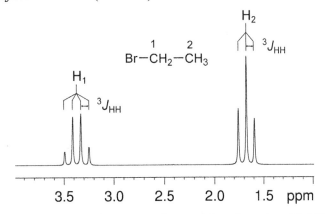

Figure 2-3   **^{1}H NMR spectrum of bromoethane (simulated at 90 MHz, CDCl$_3$) showing the multiplicity of the two ^{1}H signals.**

Nuclei which are chemically equivalent (*i.e.* have exactly the same chemical environment) do not show coupling to each other.

A signal with no splitting is commonly termed a singlet; a multiplet with two lines is termed a doublet; a multiplet with three lines is a triplet and a multiplet with four lines a quartet. For simple multiplets, the spacing between the lines (in Hz) is the coupling constant which is given the symbol "*J*". $^3J_{XY}$ indicates a coupling between nuclei X and Y through three intervening bonds.

The NMR signal of a nucleus coupled to more than one group of hydrogen atoms will be split into a multiplet-of-multiplets. In the ^{1}H NMR spectrum of 1,2,4-trichlorobenzene (Figure 2-4), the H_3 signal is split into a doublet by coupling to the *meta* proton (H_5). Similarly, the H_6 signal is split into a doublet by coupling to the *ortho* proton (H_5). The H_5 signal is split into a doublet-of-doublets by coupling to both the *ortho* and *meta* protons (H_6 and H_3, respectively).

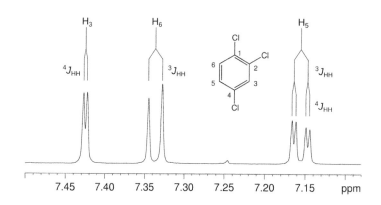

Figure 2-4 ^{1}H NMR spectrum of 1,2,4-trichlorobenzene (500 MHz, CDCl$_3$) showing the multiplicity of the three ^{1}H signals.

In saturated aliphatic systems, the two-bond coupling ($^2J_{HH}$) between two protons attached to the same tetrahedral carbon atom (*geminal* protons) typically falls in the range 10–16 Hz.

The coupling between protons attached to adjacent saturated carbons in an alkyl chain (*vicinal* protons) is typically near 7 Hz but vicinal coupling constants (H–C–C–H) in rigid saturated systems depend strongly on the dihedral angle (φ) between the two protons. $^3J_{HH}$ coupling constants in saturated systems are large for dihedral angles which approach 0° or 180°, and small for dihedral angles which are close to 90°. This relationship is known as **the Karplus relationship** (Figure 2-5). While the shape of the Karplus curve remains essentially unchanged for different molecules, the values of the constants *A*, *B* and *C* vary depending on the type of system being studied.

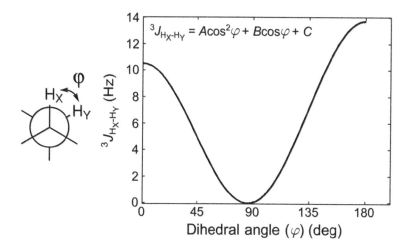

Figure 2-5 The dependence of vicinal coupling constants ($^3J_{HH}$, Hz) on dihedral angle (φ) (Karplus relationship).

In unsaturated systems, such as alkenes, vicinal coupling constants also depend on stereochemistry. Couplings between vinylic protons which are *cis* are typically in the range 6–14 Hz, while those which are *trans* are generally in the range 14–20 Hz.

In aromatic systems, *ortho*, *meta* and *para* couplings are different, and it is often possible to determine the substitution pattern of an aromatic ring simply by inspecting the number of aromatic signals and the magnitudes of the coupling constants between them.

Disubstituted benzene rings show characteristic coupling patterns, depending on the position of the substituents (Figure 2-6). Note that if the substituents are identical (*i.e.* X = Y), fewer signals will appear in the aromatic region of the spectrum.

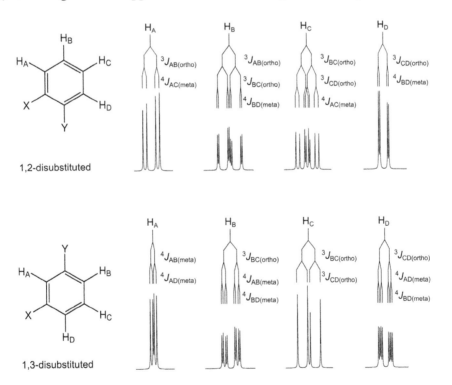

Figure 2-6 **Characteristic aromatic splitting patterns in the ^{1}H NMR spectra of some disubstituted benzene rings.**

13

The aromatic proton signals of 1,4-disubstituted benzenes often appear, superficially, as two doublets (Figure 2-7) so *para*-disubstituted benzenes are easy to identify from this characteristic coupling pattern. In fact, these signals are more complex than they initially seem, and contain many overlapping signals.

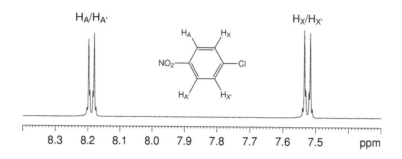

Figure 2-7 **^{1}H NMR spectrum of 1-chloro-4-nitrobenzene (500 MHz, CDCl$_3$).**

Trisubstituted benzene rings also display characteristic splitting patterns in their ^{1}H NMR spectra (Figure 2-8).

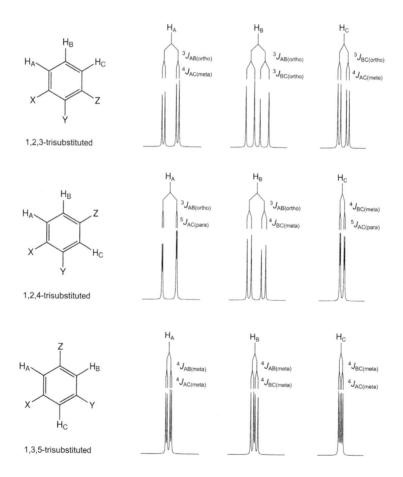

Figure 2-8 **Characteristic aromatic splitting patterns in the ^{1}H NMR spectra for some trisubstituted benzenes.**

^{1}H–^{1}H coupling is rarely observed across more than three intervening bonds unless there is a particularly favourable bonding pathway and **long-range coupling** can be observed when there is an extended π conjugation pathway or a particularly favourable rigid σ-bonding skeleton.

| $^4J_{HH}$ = 1.5 Hz | $^4J_{HH}$ = 7 Hz | $^4J_{HH}$ = 3.0 Hz | $^4J_{HH}$ = 1.25 Hz |

| $^5J_{HH}$ = 2.5 Hz | $^5J_{HH}$ = 7 Hz | $^5J_{HH}$ = 1.6 Hz |

2.2.3 *Decoupling in ^{1}H NMR Spectroscopy*

In any signal that is a multiplet due to spin–spin coupling, it is possible to remove the splitting effects by irradiating the sample with an additional Rf source at the exact resonance frequency of the nucleus giving rise to the splitting. The additional radiofrequency source causes rapid flipping of the irradiated nuclei between their available states and, as a consequence, any nuclei coupled to them cannot sense their state to cause splitting.

The process of irradiating one nucleus to remove any splitting caused by it is termed **decoupling**. Decoupling always simplifies the appearance of multiplets by removing some of the splittings. Figure 2-9 shows the ^{1}H NMR spectrum of bromoethane with decoupling of each of the ^{1}H resonances in turn. Without irradiation, H_1 appears a two-proton quartet and H_2 appears as a three-proton triplet. With irradiation at the frequency of H_1, the multiplicity of H_2 due to coupling with H_1 is removed and H_2 appears as a singlet. With irradiation at the frequency of H_2, the multiplicity of H_1 due to coupling with H_2 is removed and H_1 appears as a singlet.

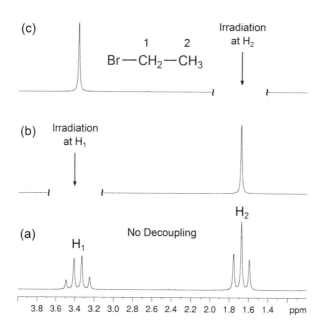

Figure 2-9 ^{1}H NMR spectrum of bromoethane (simulated at 90 MHz, CDCl₃): (*a*) basic spectrum showing all coupling; (*b*) with irradiation at H₁; (*c*) with irradiation at H₂.

2.2.4   *The Nuclear Overhauser Effect in ^{1}H NMR Spectroscopy*

Applying Rf radiation to one nucleus while observing the resonance of another may result in a change in the **amplitude** of the observed resonance, *i.e.* an enhancement or a reduction of the signal intensity. This phenomenon is known as the *nuclear Overhauser effect* (NOE). The NOE is a "through space" effect and its magnitude is inversely proportional to the sixth power of the distance between the interacting nuclei. Because of the distance dependence of the NOE, it is an important method for establishing which protons are close together in space and because the NOE can be measured quite accurately it has become a very powerful means for determining the three-dimensional structure (and stereochemistry) of organic compounds.

2.3 CARBON-13 NMR SPECTROSCOPY

The major isotope of carbon, carbon-12 or ^{12}C, is NMR-silent. However, carbon-13 or ^{13}C, which has a natural abundance of 1.1%, has a nuclear spin of ½, so NMR spectra of this isotope may be obtained easily. The low natural abundance and relatively low resonance frequency of ^{13}C means that the signals are weaker than those of ^{1}H and more spectra need to be accumulated and summed to obtain spectra where the signals are strong.

The low natural abundance of ^{13}C also means that it is very unlikely that two ^{13}C nuclei will be adjacent in a single molecule, so ^{13}C–^{13}C coupling is simply not observed in routine ^{13}C NMR spectra.

2.3.1 *Decoupling in ^{13}C NMR Spectroscopy*

^{13}C NMR spectra acquired without any ^{1}H decoupling display multiplicity due to ^{1}H–^{13}C coupling – signals from CH$_3$ carbons are split into four lines, signals from CH$_2$ carbons are split into three lines, signals from CH carbons are split into two lines and quaternary carbons appear as singlets. In practice, most ^{13}C spectra are usually recorded with **broad-band ^{1}H decoupling** – meaning that any coupling between ^{1}H and ^{13}C nuclei is eliminated by strongly irradiating *all* protons while observing the ^{13}C spectrum. This technique gives singlets (no multiplicity) for each unique carbon atom in the molecule (Figure 2-10). Such spectra are often described as ^{13}C{^{1}H} NMR spectra, where the notation of ^{1}H in curly brackets indicates that broad-band proton decoupling has been applied to ^{1}H during acquisition of the ^{13}C spectrum to remove all splitting due to C–H coupling.

While decoupling simplifies ^{13}C NMR spectra by removing the splitting from proton coupling, this means that it is not obvious which ^{13}C signals belong to CH$_3$ groups and which belong to CH$_2$ or CH groups and which carbons have no attached protons. It is always useful to know which carbon signals arise from CH$_3$ groups and those which arise from CH$_2$ groups *etc*. With most modern NMR instruments, the most commonly used 1D method to determine the multiplicity of ^{13}C signals is **the DEPT experiment** (Distortionless Enhancement by Polarisation Transfer). The DEPT experiment is a pulsed NMR experiment which requires a series of programmed Rf pulses to both the ^{1}H and ^{13}C nuclei in a sample. The resulting ^{13}C DEPT spectrum contains only signals for the protonated carbons (non-protonated carbons do not give signals in the ^{13}C DEPT spectrum). The DEPT experiment can be tuned such that signals arising from carbons in CH$_3$ and CH groups (*i.e.* those with an odd number of attached protons) appear in the opposite direction to those from CH$_2$ groups (*i.e.* those with an even number of attached protons) – signals from CH$_3$ and CH groups point upwards while signals from CH$_2$ groups point downwards (Figure 2-10).

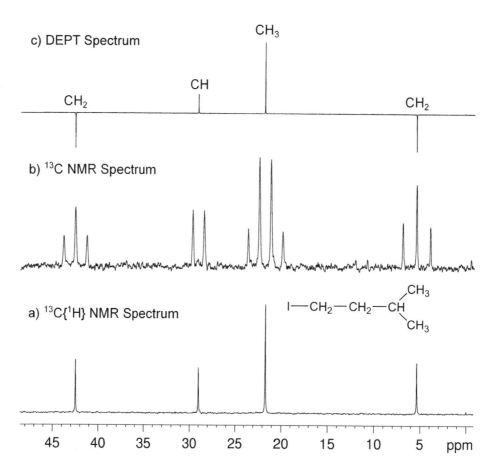

Figure 2-10   **^{13}C NMR spectra of 1-iodo-3-methylbutane (CDCl$_3$, 100 MHz): (*a*) broadband ^{1}H decoupling; (*b*) no decoupling; (*c*) DEPT spectrum.**

Coupling may also be observed between ^{13}C and other spin-active nuclei that are present in the molecule (*e.g.* ^{19}F or ^{31}P).

2.3.2   *Chemical Shifts in ^{13}C NMR Spectroscopy*

^{13}C nuclei have access to more hybridisation states (bonding geometries and electron distributions) than ^{1}H nuclei and both hybridisation and changes in electron density have a significant effect on the chemical shifts of ^{13}C nuclei. The ^{13}C chemical shift scale spans from about 0–220 ppm relative to TMS. Aliphatic, saturated carbons have chemical shifts typically in the range 10–65 ppm; aromatic carbons typically fall in the range 120–160 ppm; the sp-hybridised carbons of alkynes fall in the range 60–85 ppm and carbonyl carbons are the most deshielded and are typically found in the range 170–220 ppm. Figure 2-11 gives the chemical shift ranges for ^{13}C nuclei in common functional groups.

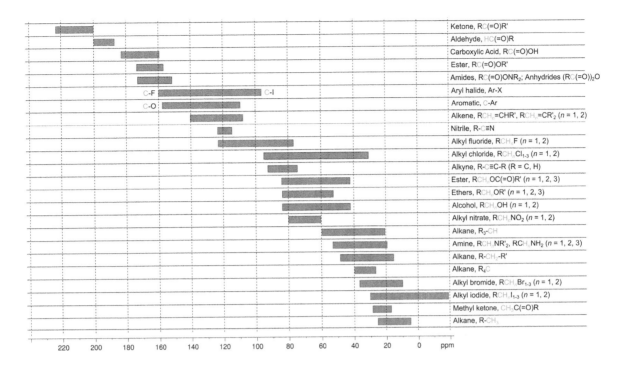

Figure 2-11   Approximate ^{13}C chemical shift ranges for carbon atoms in organic compounds.

2.4 FLUORINE-19 NMR SPECTROSCOPY

^{19}F, the single naturally occurring isotope of fluorine, is NMR-active with a nuclear spin of ½. The 100% natural abundance and high resonance frequency means that ^{19}F spectra are easily acquired. The chemical shift range for ^{19}F spectra is significantly wider than that of ^{13}C spectra, with most signals occurring in the range +200 ppm to −300 ppm (relative to CFCl$_3$ at 0 ppm). Figure 2-12 gives the chemical shift ranges for typical ^{19}F nuclei in organic molecules.

The presence of fluorine in a molecule is usually signalled by the appearance of multiplicity in ^{1}H and ^{13}C spectra due to ^{19}F–^{1}H and ^{19}F–^{13}C coupling, respectively.

Coupling between fluorine and hydrogen is strong and geminal ^{19}F–^{1}H couplings ($^2J_{FH}$) are typically around 50 Hz. Similarly, three-bond vicinal fluorine–proton coupling constants ($^3J_{FH}$) are typically between 10 and 20 Hz. Vicinal fluorine–proton couplings exhibit a Karplus-type dependence on dihedral angle, similar to that observed for vicinal proton–proton couplings (Figure 2-5).

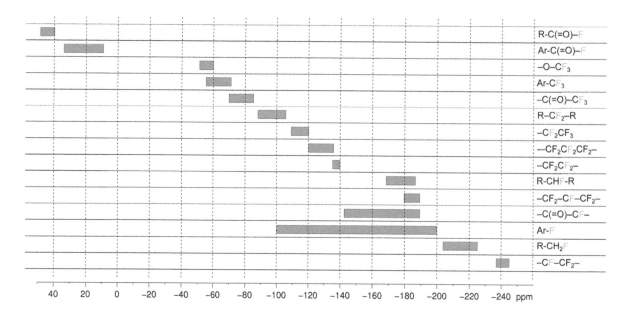

Figure 2-12   Approximate ^{19}F chemical shift ranges for fluorine atoms in organic compounds, relative to CFCl₃.

Figure 2-13 shows the ^{1}H NMR spectrum of fluoroethane. The methylene (CH_2) proton signal (H_1) is a doublet-of-quartets; the doublet splitting is due to coupling to fluorine ($^2J_{HF} = 47$ Hz) and the quartet splitting is due to proton–proton coupling between the methylene and methyl protons ($^3J_{HH} = 7$ Hz). The methyl proton signal (H_2) is a doublet of triplets where the doublet splitting is due to coupling to fluorine ($^3J_{HF} = 25$ Hz) and the triplet splitting is due to proton–proton coupling between the methylene and methyl protons ($^3J_{HH} = 7$ Hz).

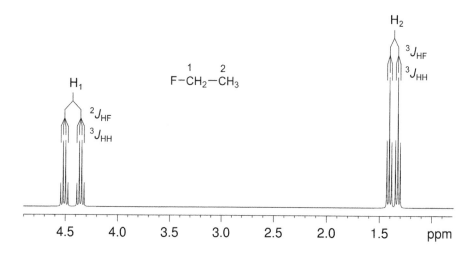

Figure 2-13 ^{1}H NMR spectrum of fluoroethane (300 MHz): $^2J_{HF} = 47$ Hz, $^3J_{HF} = 25$ Hz. Each of the ^{1}H resonances shows multiplicity due to ^{1}H–^{1}H coupling as well as a doublet splitting due to coupling to ^{19}F.

20

One-bond ^{19}F–^{13}C coupling constants ($^{1}J_{CF}$) for fluorocarbons are in the range 160–280 Hz, with increasing fluorine substitution leading to larger coupling constants (Table 2-1). The carbon–fluorine coupling constant diminishes rapidly as the carbon nuclei become further removed from the fluorine substituent – two-bond coupling constants are of the order 20–30 Hz and longer-range couplings are smaller again.

In aromatic compounds, one-bond ^{19}F–^{13}C coupling constants are typically about 240 Hz. Two-bond couplings are around 20 Hz, and couplings decrease with increasing distance from the fluorine substituent.

Table 2-1   Selected $^{n}J_{CF}$ values, in Hertz.[‡]

Compound	$^{1}J_{CF}$	$^{2}J_{CF}$	$^{3}J_{CF}$	$^{4}J_{CF}$
CH_3F	162			
CH_2F_2	234			
CHF_3	274			
CH_3CH_2F[a]	160.6	20.7		
CH_3CF_3[a]	273.0	31.5		
CH_2ICF_3	274.6	36.5		
CF_3COOH	283	43		
$CH_3CH_2CH_2CH_2F$[b]	165	19	6.0	3.1
C_6H_5F	245.3	21.0	7.7	3.3
p-CH_3–C_6H_4F	243.5	21.1	7.75	2.9

[‡] Data from Emsley, J. W.; Feeney, J.; Sutcliffe, L. H. Fluorine coupling constants. *Prog. Nucl. Mag. Res. Sp.*, **1976**, *10*, 88–756.
[a] Abraham, R. J.; Edgar, M.; Griffiths, L.; Powell, R. L., Substituent chemical shifts (SCS) in NMR. Part 5. Mono- and difluoro SCS in rigid molecules. *J. Chem. Soc., Perkin Trans. 2*, **1995**, 561–567.
[b] Dolbier, W. R., Jr. *Guide to Fluorine NMR for Organic Chemists*. John Wiley & Sons, New Jersey, **2009**.

Large carbon–fluorine couplings are easily observed in the $^{13}C\{^{1}H\}$ NMR spectra of fluorinated compounds (Figure 2-14).

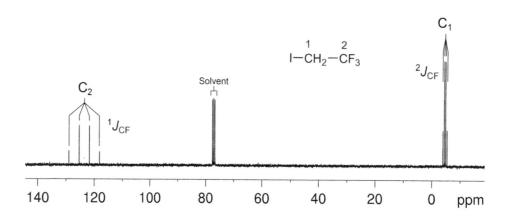

Figure 2-14 ^{13}C{^{1}H} NMR spectrum of 1-iodo-2,2,2-trifluoroethane (CDCl$_3$, 300 MHz). $^1J_{CF}$ = 274.6 Hz, $^2J_{CF}$ = 36.5 Hz.

2.5 PHOSPHORUS-31 NMR SPECTROSCOPY

^{31}P, the single naturally occurring isotope of phosphorus, is NMR-active with a nuclear spin of ½. The 100% natural abundance and relatively high resonance frequency means that ^{31}P spectra are easily acquired. Many biological compounds, as well as many organometallic transition metal complexes, contain phosphorus and ^{31}P NMR is a powerful technique used to study these types of compounds. The chemical shift range for ^{31}P spectra is significantly wider than that of ^{13}C spectra, and most signals occur in the range +200 ppm to –200 ppm (relative to 85% H$_3$PO$_4$ at 0 ppm). The oxidation state of phosphorus has a large effect on the observed ^{31}P chemical shift (Figure 2-15).

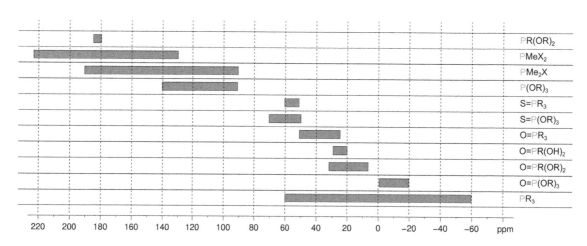

Figure 2-15 Approximate ^{31}P chemical shift ranges for phosphorus atoms in organic compounds, relative to 85% H$_3$PO$_4$.

The presence of phosphorus in a molecule is usually signalled by the appearance of multiplicity in ^{1}H and ^{13}C spectra due to ^{31}P–^{1}H and ^{31}P–^{13}C coupling, respectively. One-bond ^{31}P–^{1}H coupling constants depend on the oxidation state of phosphorus: for P^{III} compounds phosphorus–hydrogen coupling constants ($^{1}J_{PH}$) are typically between 100 and 200 Hz, while for P^{V} compounds, $^{1}J_{PH}$ are typically between 400 and 1100 Hz. For both oxidation states, $^{2}J_{PH}$ are usually between 5 and 30 Hz, and $^{3}J_{PH}$ are between 5 and 10 Hz.

The magnitude of carbon–phosphorus coupling depends on the oxidation state of phosphorus. In some cases, *e.g.* trialkylphosphine compounds (PR_3), $^{2}J_{PC}$ is often larger than $^{1}J_{PC}$.

2.6 NITROGEN-15 NMR SPECTROSCOPY

^{15}N is a rare isotope of nitrogen (~0.4%) but it is the naturally occurring isotope of nitrogen with nuclear spin of ½. ^{14}N is the major isotope of nitrogen; however, its nuclear spin is 1, *i.e.* ^{14}N is a quadrupolar nucleus and spectra for ^{14}N are typically broad and difficult to interpret. ^{15}N spectra can be obtained at natural abundance if sufficient sample is available or, more usually, the concentration of ^{15}N in a sample is enriched by chemical synthesis. The chemical shift range for ^{15}N spectra is about 1000 ppm with amines occurring near 0 ppm and nitroso compounds near 1000 ppm (Figure 2-16, relative to liquid ammonia at 0 ppm).

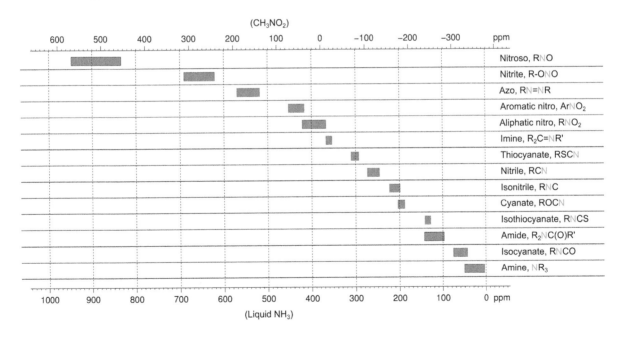

Figure 2-16 **Approximate ^{15}N chemical shift ranges for nitrogen atoms in organic compounds.**

One-bond $^{15}N-^1H$ coupling depends on the hybridisation of the N atom and the electronegativity of substituents on N. For $^{15}NH_3$, the one-bond coupling is approximately 61 Hz and for $Ph^{15}NHOH$, the one-bond coupling is approximately 79 Hz. One-bond $^{13}C-^{15}N$ couplings can be observed in organic compounds typically only where there is enrichment of the ^{15}N content; in $^{15}NMe_4^+$ $^1J_{NC}$ is approximately 6 Hz.

3 Two-Dimensional NMR Spectroscopy

3.1 GENERAL PRINCIPLES

Two-dimensional (2D) NMR spectra have signal intensity as a function of *two frequency dimensions*.

In the simplest 2D experiment, the acquisition of 2D NMR spectra involves the use of two radiofrequency (Rf) pulses, separated by an intervening time period, t_1.

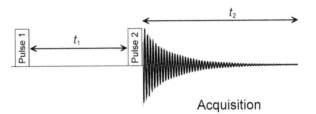

The first Rf pulse excites nuclei in the sample which can, during t_1, interact with each other through spin–spin coupling, dipolar interactions or by a range of other mechanisms. After the second pulse is applied, the free induction decay (FID) is acquired in a manner identical to that used for one-dimensional NMR spectra. The value of t_1 is then changed slightly (incremented), and the process is repeated to acquire a new FID.

In practice, the value of t_1 is systematically incremented many times (typically 512 or 1024) and separate FIDs are acquired for each value of t_1 resulting in an array of FIDs, each acquired with a different value for t_1.

Most 2D NMR experiments use more than two pulses. To generalise the 2D NMR experiment, there are four important periods in the pulse sequence used to acquire 2D NMR spectra. The *preparation sequence* excites nuclei in the sample which interact with each other through the *evolution period* t_1. The *mixing sequence* produces the magnetisation that is observed during the *detection period* t_2. The value of t_1 is then changed slightly (incremented), and the process is repeated to acquire a new FID. The preparation sequence and the mixing sequence may consist of

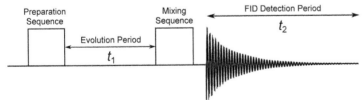

many Rf pulses separated by fixed time intervals and there are a myriad of 2D NMR experiments differing from one another by the preparation and mixing sequence they employ.

Organic Structures from 2D NMR Spectra. L. D. Field, H. L. Li and A. M. Magill
© *2015 John Wiley & Sons, Ltd. Published 2015 by John Wiley & Sons, Ltd.*

The FID for each slice in the data set is subjected to a Fourier transformation (FT) to give a new array which is effectively a series of spectra (intensity *vs.* frequency) but each spectrum in the array is different because it was acquired with a slightly different value of t_1. The array now has two dimensions, one of which is a frequency dimension (F_2) and one is still a function of time (t_1). A second FT converts the remaining time-dependent dimension to the second frequency dimension (F_1) (Figure 3-1).

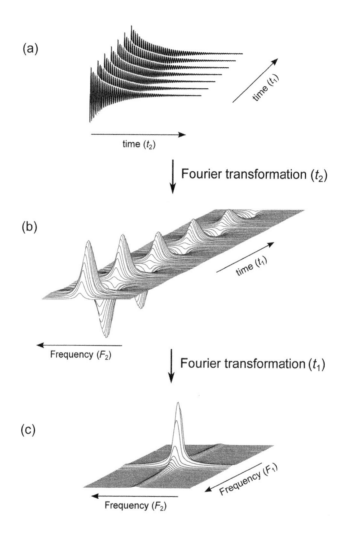

Figure 3-1   **Acquisition of a 2D NMR spectrum: (*a*) A series of individual FIDs are acquired; (*b*) Each individual FID is subjected to a Fourier transformation; (*c*) A second Fourier transformation in the remaining time dimension gives the final 2D spectrum.**

Two-dimensional NMR spectra may be displayed in several ways (Figure 3-2). The first is a type of "stacked plot", in which chemical shift in the F_2 and F_1 dimensions are represented on two separate axes and the vertical axis shows signal intensity. The second, and more common, representation of 2D NMR data is as a "contour plot", which is essentially a topographical "map" of the stacked plot. One-dimensional spectra are often shown as "projections" on the edges of contour plots, to aid interpretation of the spectra.

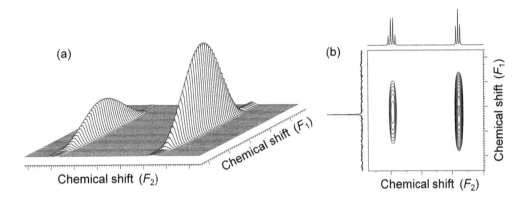

Figure 3-2   Representations of 2D NMR spectra: (*a*) Stacked plot; (*b*) Contour plot.

Two-dimensional NMR spectra may be acquired as either "magnitude" or "phase-sensitive" spectra. Magnitude spectra are obtained by taking the absolute value of each point in the 2D spectrum so, by definition, all of the peaks will be positive. Phase-sensitive spectra contain both positive and negative signals and are generally preferred, as the peak widths are narrower and, for some experiments, there is valuable information in the fine structure and relative phases of the peaks in the 2D spectrum.

When 2D NMR spectra are presented as contour plots, different colours are typically used to represent those peaks which are positive and those which are negative (Figure 3-3).

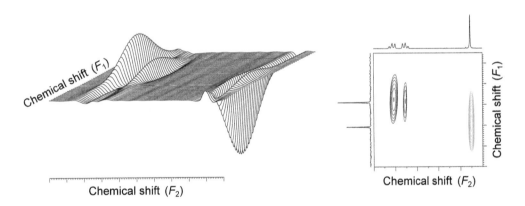

Figure 3-3   Representations of phase-sensitive 2D NMR spectra.

While, in principle, there are an unlimited number of 2D NMR experiments that can be devised, in practice, only a handful of very specific preparation and mixing sequences have emerged as essential 2D NMR experiments.

The most important 2D NMR experiments for solving structural problems are COSY (COrrelation SpectroscopY), NOESY (Nuclear Overhauser Effect SpectroscopY), HSQC (Heteronuclear Single Quantum Correlation) or HMQC (Heteronuclear Multiple Quantum Correlation), HMBC (Heteronuclear Multiple Bond Correlation), TOCSY (TOtal Correlation SpectroscopY) and INADEQUATE (Incredible Natural Abundance DoublE QUAntum Transfer Experiment).

Most modern high-field NMR spectrometers have the capability to routinely and, almost automatically, acquire COSY, NOESY, HSQC or HMQC, HMBC, TOCSY and INADEQUATE spectra.

3.2 PROTON–PROTON INTERACTIONS

3.2.1 Correlation Spectroscopy – The COSY Experiment

COSY is a homonuclear correlation technique that is mainly used to identify spin–spin coupling between protons or groups of protons (a ^1H–^1H COSY).

A COSY spectrum is always symmetrical about a diagonal in the 2D spectrum and contains peaks on the diagonal and peaks that appear off the diagonal. The symmetry in a COSY spectrum means that there is redundant information in the spectrum because, if there is a peak on one side of the diagonal, there must be a peak in the corresponding position on the other side of the diagonal.

Peaks on the diagonal occur at the frequency of each of the signals in the NMR spectrum. The off-diagonal peaks are the important signals, since these occur at positions where there is spin–spin coupling between a proton on the F_1 axis and a proton on the F_2 axis. By convention a 1D proton spectrum is usually plotted on both the F_1 and F_2 axes.

COSY is one of the simplest 2D pulse sequences. Both the preparation and mixing periods consist of a single Rf pulse. The first

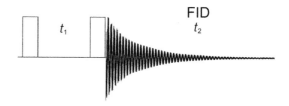

Rf pulse excites each proton in the sample and each nucleus evolves at its own resonance frequency during t_1. The second Rf pulse transfers magnetisation between spins that are coupled together so some magnetisation in the sample evolves at one frequency in t_1 and is then detected at a different frequency in t_2.

The ^{1}H–^{1}H COSY spectrum of 3-methyl-1-butanol (*iso*-amyl alcohol) is shown in Figure 3-4. The spectrum contains five peaks on the diagonal (corresponding to each unique group of protons in the molecule) and six off-diagonal cross-peaks. Because the spectrum is symmetrical, *only* those cross-peaks above the diagonal <u>or</u> those below need to be considered.

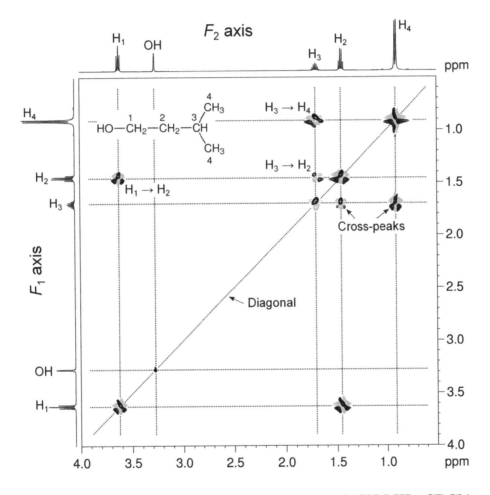

Figure 3-4 ^{1}H–^{1}H COSY spectrum of 3-methyl-1-butanol (500 MHz, CDCl$_3$).

On the basis of its chemical shift, the signal at 3.6 ppm is easily identified as H$_1$. Beginning on the F_2 axis, this signal has a single correlation to the peak at 1.45 ppm (in F_1). As cross-peaks only occur between protons which are directly coupled, the peak at 1.45 ppm in F_1 must belong to H$_2$. The H$_2$ signal has an additional cross-peak, to the signal at 1.7 ppm, which belongs to H_3. There is a cross-peak between H_3 and the last remaining unidentified peak (at 0.9 ppm), which must be due to coupling between H_3 and H_4. The OH proton does not show coupling to any of the protons in the molecule and therefore doesn't show any correlations in the COSY spectrum.

Using a COSY spectrum, it is therefore possible to sequentially identify all of the coupled spins in a single spin system, providing one can assign at least one resonance.

3.2.2   Total Correlation Spectroscopy – The TOCSY Experiment

TOCSY is a homonuclear correlation technique that identifies all the correlated spins in a spin system. Unlike the COSY experiment, which identifies only those pairs of spins which have spin–spin coupling, the TOCSY experiment correlates *all* of the spins belonging to the same spin system.

A TOCSY spectrum looks much like a COSY spectrum – it is a symmetrical spectrum with a series of peaks on the diagonal at the frequency of each of the signals in the NMR spectrum. The off-diagonal peaks (cross-peaks) occur at the position of any proton on the F_1 axis and all protons which are part of the coupled network in the F_2 dimension. A TOCSY spectrum contains a significant amount of redundant information since the whole coupled network shows up at the frequency of each individual proton in the coupled network.

The TOCSY spectrum of ethyl valerate is shown in Figure 3-5. Ethyl valerate contains two effectively independent spin systems, a butyl fragment and an ethyl fragment, separated by a carboxy (–C(=O)O–) group across which there is no proton–proton coupling. The two separate spin systems are easily identified from the TOCSY spectrum and it's obvious that one spin system has two sets of protons and the other spin system has four sets of protons. The ethyl fragment consists of a signal corresponding to H_6 at δ 3.95 ppm (assigned on the basis of its chemical shift), and a signal at δ 0.98 ppm for H_7.

The proton resonances of the butyl fragment all show correlations to each other. For example, the signal corresponding to H_4 (δ 2.1 identified on the basis of its chemical shift and multiplicity) shows three cross-peaks – one each to H_3, H_2 and H_1.

30

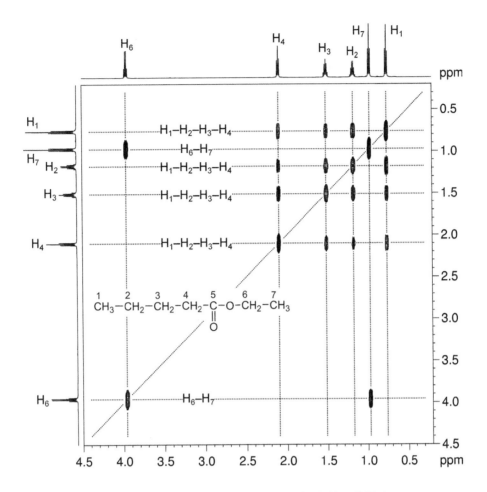

Figure 3-5   TOCSY spectrum of ethyl valerate (600 MHz, C₆D₆).

A TOCSY spectrum cannot be used to sequentially identify signals in a spin system in the same way that a COSY spectrum can; however, TOCSY is a powerful tool for identifying entire spin systems, particularly in crowded regions of the spectrum where there are several different spin systems with overlapping NMR resonances.

3.2.3   Nuclear Overhauser Spectroscopy – The NOESY Experiment

The NOESY spectrum identifies pairs of nuclei that are close together in space and relies on the nuclear Overhauser effect (NOE) (see Section 2.2.4).

¹H–¹H NOESY spectra rely on NOE interactions between protons present in the same molecule. As the NOE is only significant between nuclei which are physically close together in space, NOESY spectra can be used to determine stereochemical information about compounds, such as the relative stereochemistry around double bonds or ring junctions, the relative stereochemistry of substituents in cyclic compounds and the three-dimensional folding of biomolecules such as proteins.

^{1}H–^{1}H NOESY spectra are symmetrical about a diagonal, and peaks on the diagonal occur at the frequency of each of the signals in the NMR spectrum. The off-diagonal peaks (cross-peaks) occur at positions where there is a proton on the F_1 axis and a proton on the F_2 axis that are close together in space. By convention, a 1D proton spectrum is usually plotted on both the F_1 and F_2 axes. The intensity of the cross-peak depends on the strength of the NOE interaction, *i.e.* on how close the interacting protons are in space.

The ^{1}H–^{1}H NOESY spectrum of *trans*-ß-methylstyrene is shown below (Figure 3-6). The spectrum contains six peaks on the diagonal (corresponding to each unique group of protons in the molecule) and there are eight off-diagonal cross-peaks. Because the spectrum is symmetrical about the diagonal, *only* those cross-peaks above the diagonal <u>or</u> those below need to be considered.

The NOESY spectrum shows correlations between the methyl protons (H$_1$) and both sets of alkene protons (H$_2$ and H$_3$). There are correlations between both alkene protons and the *ortho* protons (H$_4$) of the aromatic ring. There is no correlation between the methyl protons and any proton on the aromatic ring, as these two sections of the molecule are physically too far apart.

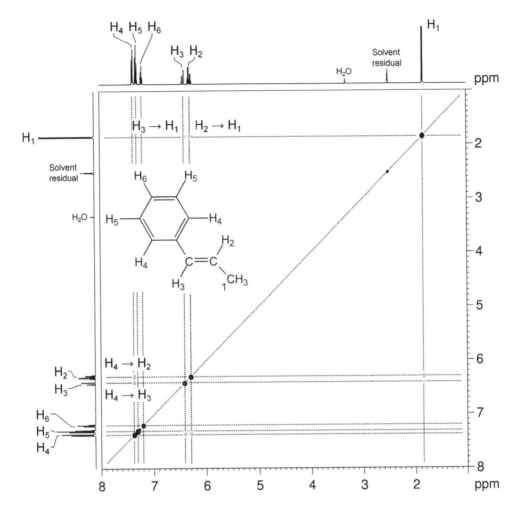

Figure 3-6 ^1H–^1H NOESY spectrum of *trans*-ß-methylstyrene (500 MHz, DMSO-d_6).
The diagonal has been plotted with reduced intensity.

Compare Figure 3-6 with the ^1H-^1H NOESY spectrum of *cis*-ß-methylstyrene (Figure 3-7). For this isomer, there are clear cross-peaks between the methyl group and the *ortho* protons of the aromatic ring ($H_4 \rightarrow H_1$), and only one of the alkene protons shows a cross-peak to the aromatic ring ($H_4 \rightarrow H_3$).

In addition, there are cross-peaks between the two alkene protons ($H_3 \rightarrow H_2$) and between one of the alkene protons and the methyl group ($H_2 \rightarrow H_1$).

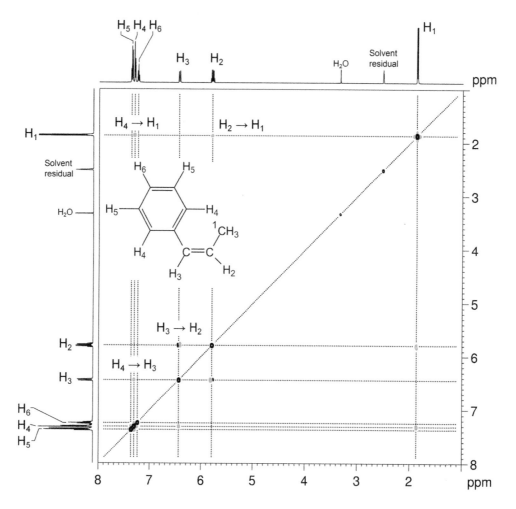

Figure 3-7 ^1H–^1H NOESY spectrum of *cis*-ß-methylstyrene (500 MHz, DMSO-d_6). The diagonal has been plotted with reduced intensity.

The cross-peaks in a NOESY spectrum may be the same phase as the diagonal, or the opposite phase. Whether NOESY cross-peaks are negative or positive depends explicitly on the rate of molecular tumbling in solution. Small molecules, like most typical small organic compounds, tumble quickly and are relatively mobile in solution, and usually give rise to negative NOESY cross-peaks relative to the diagonal. Larger molecules, like polymers or biopolymers, tumble more slowly and usually give rise to positive NOESY cross-peaks. One consequence of the change in sign of NOESY cross-peaks is that, for a certain molecular tumbling rate, there is a point where the intensity of the cross-peaks may be effectively zero or vanishingly small.

3.3 CARBON–CARBON INTERACTIONS

3.3.1 The INADEQUATE Experiment

The INADEQUATE spectrum identifies pairs of carbon nuclei that are adjacent to one another.

It is always difficult to detect ^{13}C–^{13}C connectivity by NMR because the low natural abundance ^{13}C (~1.1%) means that the probability of finding two ^{13}C nuclei adjacent to each other in the same molecule is very low. The INADEQUATE experiment is designed to detect those rare instances where two ^{13}C atoms are adjacent and couple to each other and this allows correlation between them.

Until recently, the INADEQUATE experiment was impractical because the amount of sample required to obtain spectra was prohibitively large. However, in recent years, with advent of very sensitive cryogenically cooled probes for ^{13}C NMR spectroscopy, the INADEQUATE experiment is now more routinely available. The INADEQUATE experiment directly indicates which carbons are adjacent to each other in a molecule so the carbon skeleton of an unknown compound can be directly established.

The appearance of an INADEQUATE spectrum is quite different to that of other 2D correlation spectra. A 1D ^{13}C spectrum is plotted along the F_2 axis to give a reference spectrum. The F_1 dimension is a frequency dimension which represents the sum of the chemical shifts of the ^{13}C nuclei which are correlated. The relative positions of the peaks in the F_1 dimension do not play a role in interpreting the INADEQUATE spectrum.

Correlations between adjacent carbon nuclei appear as pairs of peaks on the same horizontal level and each individual peak is a doublet, with the splitting equal to $^1J_{CC}$. The spectrum does not have a diagonal corner-to-corner in the usual sense, but there is a "pseudo-diagonal", which runs through the midpoints of each pair of correlated peaks. There are no peaks on the pseudo-diagonal; however, each pair of correlated peaks is symmetrically disposed around the pseudo-diagonal.

The INADEQUATE spectrum of 2-methyl-1-butanol is shown in Figure 3-8. The spectrum shows four pairs of correlated peaks with each pair symmetrically disposed around the pseudo-diagonal.

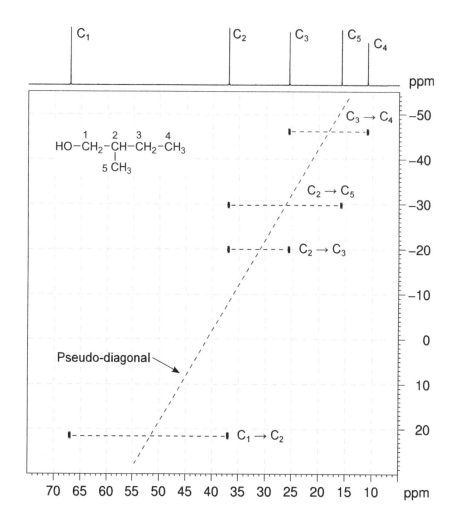

Figure 3-8   INADEQUATE spectrum of 2-methyl-1-butanol (125 MHz, CDCl₃).

The low-field chemical shift of the signal at 67 ppm identifies it as C_1, the carbon bearing the –OH substituent. The C_1 resonance has only one cross-peak in the spectrum at the same frequency in F_1 and this must correspond to C_2 (at 37 ppm). C_2 must be a branch point in the carbon chain since it shows correlations to two additional cross-peaks, which correspond to C_3 (at 26 ppm) and C_5 (at 17 ppm). C_3 has one additional cross-peak which identifies C_4 (at 11 ppm). Interpreting an INADEQUATE spectrum is a matter of taking sequential horizontal and vertical steps on the 2D spectrum until the full carbon skeleton of the molecule is revealed.

When there is sufficient sample available to make a solution of high concentration and when instruments are available to record ¹³C NMR at high sensitivity, the INADEQUATE experiment is arguably the most powerful and most useful NMR experiment for identifying organic structures.

3.4 HETERONUCLEAR CORRELATION SPECTROSCOPY

3.4.1 *Heteronuclear Single Bond Correlation – The HSQC, HMQC and me-HSQC Experiments*

HSQC, HMQC and me-HSQC spectra correlate the protons in a molecule with the carbon atoms *to which they are directly attached.*

While there are technical differences in the pulse sequences that the spectrometer uses to acquire the spectra, in practice, the HSQC and HMQC experiments provide essentially the same information. The multiplicity-edited HSQC provides additional information about whether the signals arise from CH, CH_2 or CH_3 groups, in much the same way as a 1D DEPT (Distortionless Enhancement by Polarisation Transfer) experiment (see Section 2.3.1). In the *me*-HSQC spectrum, signals from the $-CH_2-$ carbons appear with the opposite phase to those of $-CH_3$ and $-CH-$ carbons.

Heteronuclear correlation spectra are typically plotted with a 1D ^1H NMR spectrum along the proton (F_2) axis, and a 1D $^{13}C\{^1H\}$ NMR spectrum along the carbon (F_1) axis to provide a reference for the cross-peaks that occur in the 2D spectrum.

Cross-peaks in the HSQC, HMQC and me-HSQC experiments occur at the chemical shift of a proton resonance in F_2 and the chemical shift of the carbon to which it is directly bound in F_1. The HSQC, HMQC and me-HSQC experiments all rely on the large coupling between a ^{13}C and its attached proton(s). HSQC, HMQC and me-HSQC spectra are all tuned to identify correlations that occur through a single bond only, *i.e.* to detect only the correlations that arise from directly bound protons. Non-protonated carbons *do not* give rise to cross-peaks in HSQC, HMQC and me-HSQC spectra.

The multiplicity-edited HSQC spectrum of 3-methyl-1-butanol (*iso*-amyl alcohol) is shown in Figure 3-9. Unlike the homonuclear ^1H–^1H correlation experiments, such as COSY, TOCSY and NOESY, the heteronuclear correlation experiments cannot contain a diagonal.

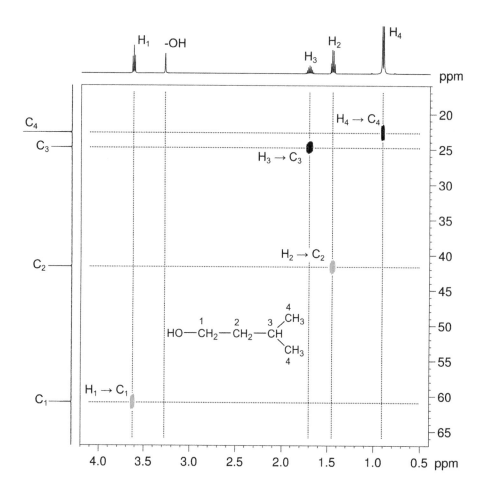

Figure 3-9   Multiplicity-edited HSQC spectrum of 3-methyl-1-butanol (500 MHz, CDCl₃). Positive contours (CH/CH₃) are shown in black and negative contours (CH₂) in red.

3-Methyl-1-butanol has four carbon and five proton resonances, and the me-HSQC shows four cross-peaks. Having used the ¹H–¹H COSY spectrum (see Figure 3-4) to identify each of the resonances in the ¹H NMR spectrum, the correlations in the me-HSQC spectrum may be used to identify each carbon resonance. Note that, because this is a multiplicity-edited HSQC, the CH₂ carbons (C₁ and C₂) appear negative (red contours) and the CH and CH₃ carbons (C₃ and C₄) are phased positive (black contours). There are no correlations from the –OH proton because this proton is not directly bound to a carbon atom.

3.4.2   Heteronuclear Multiple Bond Correlation – HMBC

HMBC spectra correlate protons with nearby carbon atoms, *not* to the carbon to which the proton is directly attached. The HMBC spectrum is normally plotted with a 1D proton spectrum on the F_2 axis and a 1D carbon spectrum on the F_1 axis, to provide a reference for the cross-peaks that are observed in the 2D spectrum.

The HMBC experiment relies on the long-range coupling between a ^{13}C and protons which are nearby. The HMBC spectra are all tuned to supress correlations that occur through a single bond and to detect only the correlations that arise from more remote protons – typically from protons two or three bonds away.

Because correlations can occur through quaternary carbon atoms, and through bridging heteroatoms such as oxygen, nitrogen or sulfur, the HMBC experiment is a powerful way to link together separate fragments of a molecule that might otherwise be difficult to connect.

The ^1H–^{13}C HMBC spectrum of ethyl valerate is shown in Figure 3-10. This compound has six proton resonances and seven carbon resonances, and the HMBC spectrum shows 15 cross-peaks. H_1 correlates to C_2 (a two-bond correlation) and C_3 (a three-bond correlation). H_2 correlates to C_1, C_3 (two-bond correlations) and C_4 (a three-bond correlation); H_3 correlates to C_2, C_4 (two-bond correlations), C_1 and C_5 (three-bond correlations); H_4 correlates to C_3, C_5 (two-bond correlations) and C_2 (a three-bond correlation); H_6 correlates to C_7 (a two-bond correlation) and to C_5 (a three-bond correlation); and H_7 correlates to C_6 (a two-bond correlation). In general, a four-bond separation is too large to permit an HMBC correlation however in some rigid structures, where there are particularly favourable long-range coupling pathways, HMBC correlations through more than two or three bonds can be observed.

Note that in ethyl valerate there is a correlation from H_6 to C_5 that bridges across the ester oxygen and this clearly links the two otherwise separate sections of the molecule.

Two-bond correlations

Three-bond correlations

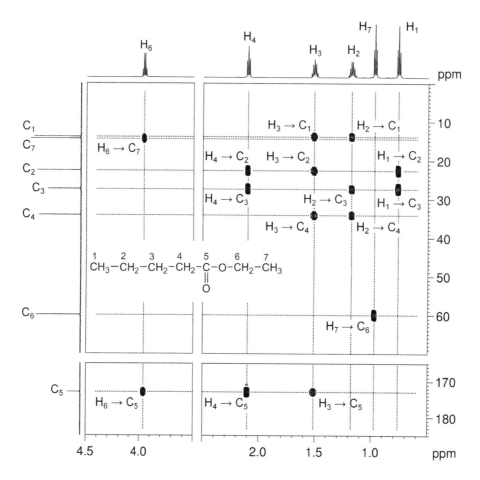

Figure 3-10 ^{1}H – ^{13}C HMBC spectrum of ethyl valerate (600 MHz, CDCl$_3$). **Portions of the spectrum which do not contain any information have been removed for clarity. The carbon resonances were assigned using a combination of ^{1}H–^{1}H COSY and ^{1}H–^{13}C me-HSQC spectra.**

HMBC spectra in aromatic systems. In aromatic systems, it is generally found that $^3J_{CH} > {}^2J_{CH} \approx {}^4J_{CH}$ (Table 3-1), so HMBC spectra of aromatic systems typically show strong cross-peaks between ^{1}H and ^{13}C that are separated by three bonds (*i.e.* between a proton and its *meta* carbons), while cross-peaks between ^{1}H and ^{13}C that are separated by two bonds (*i.e.* between a proton and its *ortho* carbons) or four bonds (*i.e.* between a proton and its *para* carbon) are usually very weak, or may often be entirely absent from the spectrum.

Table 3-1 One-, two-, three- and four-bond C–H coupling constants in benzene and some mono-substituted benzenes.

X	$^1J_{C_1-H_1}$	$^2J_{C_1-H_2}$	$^3J_{C_1-H_3}$	$^4J_{C_1-H_4}$
H	159.0	1.0	7.6	−1.2
CH$_3$		0.5	7.6	−1.4
Cl		−3.4	10.9	−1.8
Br		−3.3	11.2	−1.9
OH		−2.8	9.7	−1.6
CHO		0.3	7.2	−1.3

The HMBC spectrum of 1-bromo-2-chlorobenzene (Figure 3-11) shows eight strong cross-peaks resulting from three-bond correlations (H$_3$ to C$_1$ and C$_5$; H$_4$ to C$_2$ and C$_6$; H$_5$ to C$_1$ and C$_3$; and H$_6$ to C$_2$ and C$_4$). There are two very weak cross-peaks resulting from two-bond correlations (H$_3$ to C$_2$ and H$_6$ to C$_1$). No other two-bond correlations are present in the spectrum.

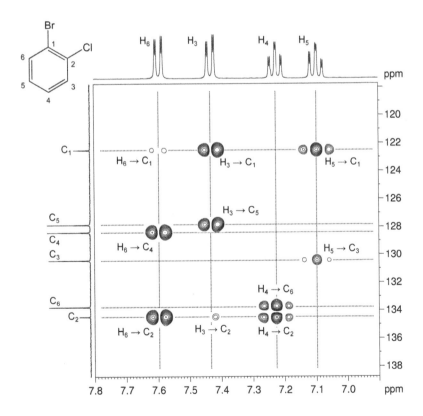

Figure 3-11 ^{1}H–^{13}C HMBC spectrum of 1-bromo-2-chlorobenzene (400 MHz, CDCl$_3$).

Three-bond correlations also occur between aromatic protons and carbon atoms directly bound to an aromatic ring, for example in toluene, benzonitrile or benzoic acid. The α-protons of ring substituents, such as the methyl protons of toluene, or the aldehydic proton of benzaldehyde, usually show both two-bond and three-bond correlations to carbons of the aromatic ring. The HMBC spectrum is as a powerful tool to locate the positions of substituents on a poly-substituted aromatic ring.

$^3J_{CH} = 4.6$ Hz $^3J_{CH} = 4.9$ Hz $^3J_{CH} = 4.1$ Hz

$^2J_{CH} = -6.0$ Hz $^2J_{CH} = 24.1$ Hz
$^3J_{CH} = 5.0$ Hz $^3J_{CH} = 2.1$ Hz

The effect of symmetry on the HMBC spectra of aromatic compounds.   If an aromatic ring contains a mirror plane that passes through two carbon atoms in the ring, additional peaks will be observed in the HMBC spectrum. While these peaks appear to arise from one-bond correlations, they arise from three-bond correlations between the proton and the symmetry-related *meta* carbon.

The HSQC and HMBC spectra of iodobenzene (Figure 3-12) illustrate this effect clearly. From the HSQC spectrum, the direct one-bond correlations between the proton and carbon resonances may be identified. The HMBC spectra contain the usual long-range correlations but also two

Mono-substituted Benzene   Di-substituted Benzenes

Tri-substituted Benzenes   Tetra-substituted Benzene

correlations which appear to be one-bond correlation between H_2 and $C_{2'}$, and H_3 and $C_{3'}$.

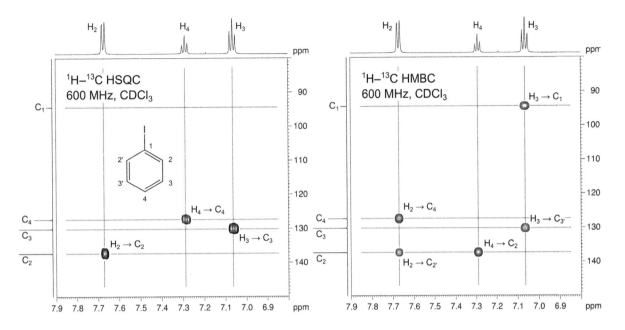

Figure 3-12 ^{1}H–^{13}C HSQC and HMBC spectra of iodobenzene (600 MHz, CDCl$_3$).

Apparent one-bond correlations are also observed in the HMBC spectra of *tert*-butyl, isopropyl and *gem*-dimethyl compounds. Again, these correlations arise from the $^3J_{CH}$ correlation between the protons of one of the methyl groups with the chemically equivalent carbon which is three bonds away.

One-bond artefacts in HMBC spectra. The HMBC experiment suppresses large coupling constants which typically correspond to one-bond correlations. In some cases however, this suppression is incomplete. Most commonly, incomplete one-bond suppression occurs in cases where the one-bond coupling constant, $^1J_{CH}$, is much larger than average. Typically, $^1J_{CH}$ values fall between 120 and 160 Hz, but some compounds such as alkynes ($^1J_{CH} \approx 250$ Hz) and aldehydes ($^1J_{CH} \approx 170$ Hz) have values that fall outside of this range and are not completely suppressed.

The ^{1}H–^{13}C HMBC spectrum of 1,1,2-trichloroethane is shown in Figure 3-13. This spectrum shows the two expected cross-peaks corresponding to H$_1$ to C$_2$ and H$_2$ to C$_1$ correlations, but H$_1$ shows a further correlation peak which aligns with C$_1$. This peak is a doublet, with a large peak separation equal to $^1J_{CH}$ (182 Hz). Doublet peaks with a large splitting are easily identified and are usually due to unsuppressed one-bond correlations.

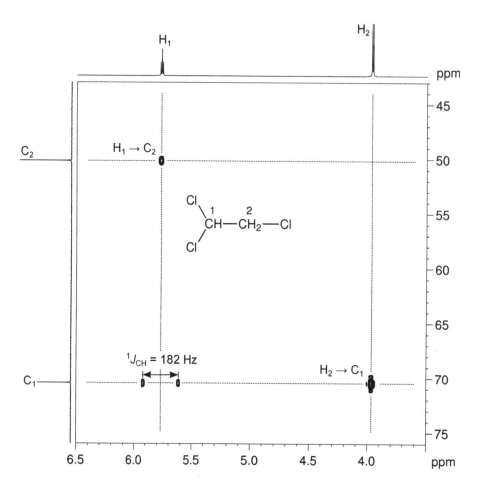

Figure 3-13 ^1H–^{13}C HMBC spectrum of 1,1,2-trichloroethane (600 MHz, CDCl₃). The unsuppressed one-bond correlation between C_1 and H_1 is seen as a doublet, the peak separation of which is equal to $^1J_{CH}$.

4 Miscellaneous Topics

4.1 NMR SOLVENTS

The choice of which NMR solvent to use for a sample depends on a number of important factors, in particular:

i. *Solubility.* Compounds need to be soluble in the chosen solvent, but not necessarily *very* soluble because modern NMR spectrometers can routinely acquire spectra with an adequate signal-to-noise ratio using as little as 1 mg of compound in ~0.7 mL of deuterated solvent.

ii. *Position of solvent signals.* In order to reduce the size of the solvent signal(s) relative to the solute signal(s), most NMR solvents are deuterated, *i.e.* have all ^1H nuclei replaced with deuterium (^2H). All deuterated solvents display some "solvent residual" signals in the ^1H NMR spectra, which are ^1H resonances due to incompletely deuterated solvent molecules. Most solvents also give signals in ^{13}C spectra. As much as possible, the solvent should be chosen to minimise the overlap of solute resonances with signals arising from the solvent.

iii. *Price.* Some NMR solvents are cheaper than others. Deuterated chloroform ($CDCl_3$) and deuterium oxide (D_2O) are relatively inexpensive and used extensively, while more exotic and more expensive solvents such as deuterated tetrahydrofuran (THF-d_8) are typically only used when other solvents are unsuitable.

iv. *Solvent-induced shifts.* Generally solvents chosen for NMR spectroscopy do not associate strongly with the solute. However, in some instances, solvents are chosen deliberately to associate with the solute and to influence chemical shifts. Changing solvents is a useful technique to move some chemical shifts around and to remove overlap from crowded regions of the spectrum. The most useful solvents for solvent-induced shifts are aromatic solvents, in particular hexadeuterobenzene (C_6D_6).

Many NMR solvents are hygroscopic (absorb moisture from the surrounding atmosphere) and their spectra may also contain small signals due to the presence of H_2O.

Table 4-1 gives the chemicals shifts of the residual signals from solvents which are commonly used in NMR spectroscopy.

Organic Structures from 2D NMR Spectra. L. D. Field, H. L. Li and A. M. Magill
© *2015 John Wiley & Sons, Ltd. Published 2015 by John Wiley & Sons, Ltd.*

Table 4-1 ^1H and ^{13}C chemical shifts for common NMR solvents, and ^1H chemical shift of residual water.

Solvent	^1H Chemical Shift (multiplicity)	^{13}C Chemical Shift (multiplicity)	^1H Chemical shift of H_2O
Acetone-d_6	2.05 (5)	29.92 (7)	2.8
		206.68 (1)	
Acetonitrile-d_3	1.94 (5)	1.39 (7)	2.1
		118.69 (1)	
Benzene-d_6	7.16 (1)	128.39 (3)	0.4
Chloroform-d	7.24 (1)	77.23 (3)	1.5
Deuterium oxide	4.80 (1)	—	4.8
Dichloromethane-d_2	5.32 (3)	54.00 (5)	1.5
Dimethyl sulfoxide-d_6	2.50 (5)	39.51 (7)	3.3
Methanol-d_4	3.31 (5)	49.15 (7)	—[a]
	4.78 (1)		
Tetrahydrofuran-d_8	1.73 (1)	25.37 (5)	2.4-2.5
	3.58 (1)	67.57 (5)	
Toluene-d_8	2.09 (5)	20.4 (7)	0.4
	6.98 (5)	125.49 (3)	
	7.00 (1)	128.33 (3)	
	7.09 (m)	129.24 (3)	
		137.86 (1)	

[a] Rapidly exchanges with the –OD of methanol.

Most NMR solvents are deuterated, and coupling to deuterium is usually observed in both the ^1H and ^{13}C NMR spectra of these solvents. Recall that deuterium has a nuclear spin of 1, so coupling patterns differ from those normally observed between two (or more) spin ½ nuclei (Figure 4-1). The patterns result from the coupling of ^{13}C to one, two or three deuterium atoms.

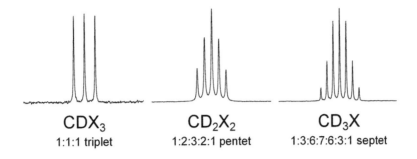

CDX₃	CD₂X₂	CD₃X

Figure 4-1   Common coupling patterns for solvent signals in ^{13}C spectra.

4.2 REFERENCE COMPOUNDS AND STANDARDS

All NMR spectra are referenced to a known compound whose chemical shift is defined to be 0 ppm. For ^{1}H and ^{13}C NMR spectra, this compound is tetramethylsilane (TMS). When recording NMR spectra, it is common to add the reference to the sample to act as an *internal standard*. Ideally, an internal standard should:

i.   not react with either the solvent or the solute;

ii.   have a single, sharp signal that is well separated from the signals of the solvent and solute;

iii. have a chemical shift that doesn't depend strongly on concentration, temperature, solvent, *etc.*;

iv.   be easily removable from the sample if required;

v.   not be highly toxic.

TMS is a suitable reference compound for proton and carbon spectra as it is chemically inert, resonates at a chemical shift that is removed from the majority of organic compounds, and is volatile (b.p. 25.6°C). For aqueous samples, 2,2-dimethyl-2-silapentane-5-sulfonate (DSS) or trimethylsilyl propionate (TSP) (or the sodium salts of these compounds) are appropriate substitutes for TMS.

Some reference compounds, such as 85% H_3PO_4, which is used as a reference compound for ^{31}P NMR spectroscopy, are too reactive to be used as internal standards. In such cases, spectra may be run by placing a separate axial capillary inside the sample solution, or using a NMR tube with an outer, concentric jacket containing the reference compound.

Table 4-2 gives the common reference compounds for common nuclei observed by NMR.

Table 4-2   Reference compounds used in NMR spectroscopy.

Isotope	Reference Compound	Conditions
^1H	$Si(CH_3)_4$	1% in $CDCl_3$
	DSS Methyl Signal	D_2O
^{13}C	$Si(CH_3)_4$	1% in $CDCl_3$
	DSS Methyl Signal	D_2O
$^{15}N^a$	$MeNO_2$	90% in $CDCl_3$
	NH_3	Neat liquid
^{19}F	$CFCl_3$	
^{31}P	H_3PO_4	External Reference

^{a} $MeNO_2$ has a shift 380 ppm to low field of NH_3.

4.3 DYNAMIC PROCESSES

The position of an NMR signal depends on the chemical environment of the resonating nucleus. In most cases, the chemical environment of a nucleus does not change during the acquisition of an NMR spectrum. However, there are instances where the chemical environment of a nucleus changes during the course of acquisition. As a result, the chemical shift of the nucleus changes during the NMR experiment.

The exchange of a nucleus between one environment and another is common in molecules where there is *(i)* intermolecular exchange where a proton is labile and exchanges with a solvent or with a proton in another molecule; or *(ii)* intramolecular exchange such as an internal conformational change, *e.g.* chair-boat-chair ring inversions in cyclic compounds or hindered rotation around some bonds.

When the rate of exchange (k, sec^{-1}) between two magnetically distinct nuclei which are exchanging with each other is very slow compared to the difference in their chemical shifts (Δv, Hz), the signals for each nucleus are separate in an NMR spectrum (Figure 4-2). As the rate of exchange increases, the signals broaden until a coalescence point is reached – here the two nuclei can no longer be distinguished by NMR spectroscopy. If the rate of exchange increases further (fast exchange) the two nuclei will be observed as a single signal at an intermediate (averaged) frequency.

The most common method to increase and decrease the rate of an exchange process is to alter the temperature of the solution – increasing temperature will increase the rate of exchange and *vice versa*.

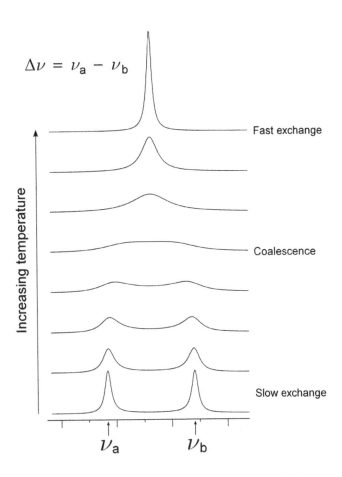

$$\Delta\nu = \nu_a - \nu_b$$

Fast exchange

Coalescence

Slow exchange

ν_a ν_b

Increasing temperature

Figure 4-2   Schematic NMR spectra of two exchanging nuclei.

4.3.1   Protons on Heteroatoms

Protons in alcohols (–OH), carboxylic acids (–COOH), primary and secondary amines (–NHR) and thiols (–SH) are loosely bound and undergo rapid exchange, either with each other or with the protons of any residual water remaining in the solvent. Rapid exchange means that the protons are not usually bound to the heteroatom long enough to couple to any neighbouring protons, so signals due to exchangeable protons typically appear as broadened singlets.

Some deuterated NMR solvents (such as D_2O or methanol-d_4) contain exchangeable deuterons and may exchange their deuterium atoms with the solute under investigation if the solute contains exchangeable protons. H/D exchange is a useful technique to identify the exchangeable protons in a compound. ^1H NMR spectra are acquired before and after the addition of a drop of D_2O to the sample, any resonances which disappear from the ^1H NMR spectrum after addition of a drop of D_2O belong to exchangeable protons in the sample.

The NH protons in amides (–C(=O)–NHR) exchange only slowly with D_2O and exchange often requires heating or the addition of a catalytic amount of base. Coupling ($^3J_{HH}$) is sometimes observed between the amide protons and protons on adjacent carbon atoms.

4.3.2 *Rotation about Partial Double Bonds*

Amides often show evidence of restricted rotation around the C–N bond of the amide functional group. While the C–N bond is formally a single bond where one might expect free rotation, the C–N bond has "*partial double bond character*" which can be rationalised by the strong contribution of a stable resonance structure that contains a C=N double bond.

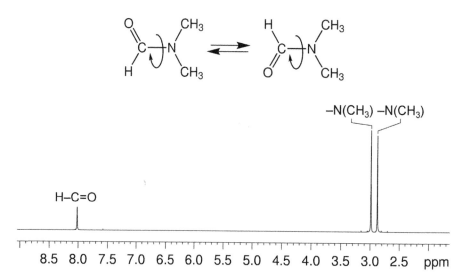

Figure 4-3 shows the ^1H NMR spectrum of *N,N*-dimethylformamide. In this case, it is clear that there are two separate signals corresponding to the CH_3 groups of *N,N*-dimethylformamide and this indicates that there is restricted rotation about the C–N bond and one of the methyl groups is *cis* to the formamide proton and one is *trans*. If the temperature is raised, the signals for the methyl groups collapse to a single averaged resonance as the rate of rotation about the C–N bond increases.

Figure 4-3 ^1H **NMR spectrum of *N,N*-dimethylformamide (CDCl₃, 400 MHz).**

4.4 SECOND-ORDER EFFECTS

Providing the difference between the chemical shifts of coupled protons (Δv) is much larger than the value of the coupling constant (J) between them, NMR resonances appear as simple singlets, doublets, triplets, *etc.* Such spectra are called "first-order spectra" and the chemical shift and coupling constant information can be measured directly from the multiplets in the spectra. As the ratio $\Delta v/J$ decreases, multiplets become skewed and distorted until eventually chemical shift and coupling constant information can no longer be measured directly from the spectrum (Figure 4-4) and such spectra are termed "second-order spectra".

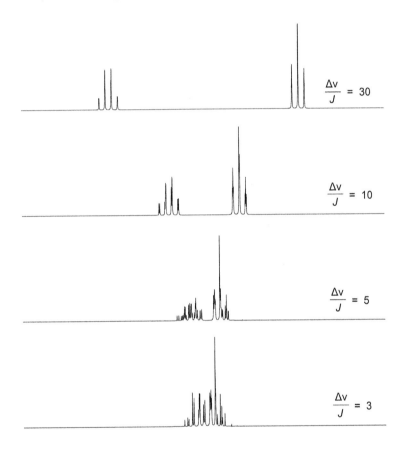

Figure 4-4   **Simulated ^{1}H NMR spectra of coupled CH$_3$ and CH$_2$ groups as the ratio $\Delta v/J$ is varied.**

4.5 EFFECT OF A CHIRAL CENTRE ON NMR SPECTRA

The presence of a chiral centre in a molecule has the effect of rendering the protons attached to a prochiral centre non-equivalent. A prochiral centre is most commonly an sp^3 hybridised carbon atom with two identical substituents and two unique substituents (R$_1$–CX$_2$–R$_2$), *e.g.* an R$_1$–CH$_2$–R$_2$ group or an R$_1$–C(CH$_3$)$_2$–R$_2$ group. Whenever a molecule contains a chiral centre, the protons of each –CH$_2$– group will be non-equivalent and they are termed diastereotopic protons.

The ^{1}H NMR signals for the CH and CH$_2$ protons of 1-phenyl-1-hydroxy-4,4-dimethylpentan-3-one are shown in Figure 4-5. Three unique signals are present, and separate resonances are observed for H$_A$ and H$_B$ even though these protons are bound to the same carbon.

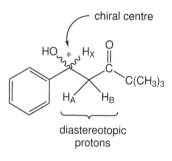

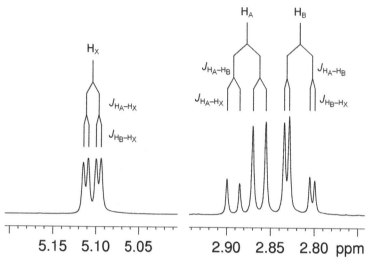

Figure 4-5   **Methylene (CH$_2$) and methine (CH) ^{1}H NMR signals of 1-phenyl-1-hydroxy-4,4-dimethylpentan-3-one (600 MHz, CDCl$_3$).**

H$_A$ and H$_B$ have distinct chemical shifts and are coupled firstly to each other ($^2J_{HH} \approx 17$ Hz) and then H$_A$ and H$_B$ each show coupling to H$_X$.

The non-equivalence of protons attached to a prochiral centre in the presence of a chiral centre can be easily rationalised. The Newman projections for three possible rotamers of 1-phenyl-1-hydroxy-4,4-dimethylpentan-3-one are shown below (R = C(=O)C(CH$_3$)$_3$). In rotamer III where H$_A$ sits between the –Ph and the –OH substituents, its magnetic environment is different to that of H$_B$ when it sits between the –Ph and the –OH substituents (rotamer I). In each of the rotamers I, II and III, the chemical environments of H$_A$ and H$_B$ are different, so an average shift of H$_A$ (over all conformations) will be different to the average shift for H$_B$.

The effect of a chiral centre extends to groups other than –CH$_2$– groups. Isopropyl groups or other –CR$_2$– groups also have non-equivalent "R" groups when there is a chiral centre in the molecule. In the ^{1}H NMR spectrum for 1-phenyl-2-methylpropanol (Figure 4-6), the methyl groups give rise to separate signals.

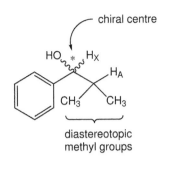

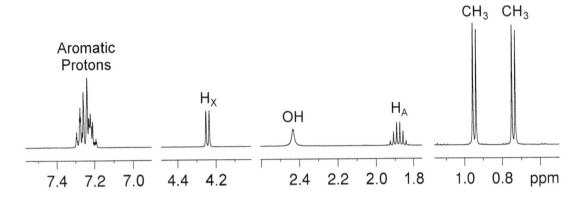

Figure 4-6 ^{1}H NMR spectrum of 1-phenyl-2-methyl-1-propanol (400 MHz, CDCl$_3$).

5 Worked Examples

5.1 GENERAL PRINCIPLES

The best way to develop expertise in solving "organic structures from spectra" is to practise. It should also be emphasised that there are many paths to the correct answer – there is no single process to arrive at the correct solution to any problem.

In this collection of problems, we have assumed a working knowledge of basic structural organic chemistry and common functional groups. We also assume a working knowledge of the rudimentary spectroscopic methods which would be applied routinely in characterising and identifying organic compounds including infrared spectroscopy and basic 1D ^{13}C and ^{1}H NMR spectroscopy.

When presented with a problem containing some basic characterisation data and 2D NMR data, students often find the following approach useful:

i. Extract as much information as possible from the basic characterisation data which is available:

 a. **Note the molecular formula** and any restrictions this places on the functional groups that may be contained in the molecule.

 b. From the molecular formula, **determine the degree of unsaturation.** The degree of unsaturation can be calculated from the molecular formula for all compounds containing C, H, N, O, S and the halogens using the following three basic steps:

 1. Take the molecular formula and replace all halogens by hydrogens;

 2. Omit all of the sulfur and/or oxygen atoms;

 3. For each nitrogen, omit the nitrogen and omit one hydrogen.

 After these three steps, the molecular formula is reduced to C_nH_m, and the degree of unsaturation is given by

$$\text{Degree of unsaturation} \quad = \quad n - {}^m/_2 + 1$$

 The degree of unsaturation indicates the number of π bonds and/or rings that the compound contains. For example if the degree of unsaturation is 1, the molecule can only contain one double bond or one ring. If the degree of

Organic Structures from 2D NMR Spectra. L. D. Field, H. L. Li and A. M. Magill
© 2015 John Wiley & Sons, Ltd. Published 2015 by John Wiley & Sons, Ltd.

unsaturation is 4, the molecule must contain four rings or multiple bonds. An aromatic ring accounts for four degrees of unsaturation (the equivalent of three double bonds and a ring). An alkyne or a C≡N accounts for two degrees of unsaturation (the equivalent of two π bonds).

c. **Analyse the 1D ¹H NMR spectrum** if one is provided and note the relative numbers of protons in different environments and any obvious information contained in the coupling patterns. Note the presence of aromatic protons, exchangeable protons, and/or vinylic protons, all of which provide valuable information on the functional groups which may be present.

d. **Analyse the 1D ¹³C NMR spectrum** if one is provided and note the number of carbons in different environments. Note also any resonances that would be characteristic of specific functional groups, *e.g.* the presence or absence of a ketone, aldehyde, ester or carboxylic acid carbonyl resonance.

e. **Analyse any infrared data** and note whether there are absorptions characteristic of specific functional groups, *e.g.* C=O or –OH groups.

ii. **Extract basic information from the 2D COSY, TOCSY and/or C–H correlation spectra**.

a. The COSY will provide obvious coupling partners. If there is one identifiable starting point in a spin system, the COSY will allow the successive identification (*i.e.* the sequence) of all nuclei in the spin system. The COSY cannot jump across breaks in the spin system (such as where there is a heteroatom or a carbonyl group that isolates one spin system from another).

b. The TOCSY identifies all groups of protons that are in the same spin system.

c. The C–H correlation links the carbon signals with their attached protons and also identifies how many –CH–, –CH₂–, –CH₃ and quaternary carbons are in the molecule.

iii. **Analyse the INADEQUATE spectrum** if one is provided, because this can sequentially provide the whole carbon skeleton of the molecule. Choose one signal as a starting point and sequentially work through the INADEQUATE spectrum to determine which carbons are connected to which.

iv. **Analyse the HMBC spectrum.** This is perhaps the most useful technique to pull together all of the fragments of a molecule because it gives long-range connectivity.

v. **Analyse the NOESY spectrum** to assign any stereochemistry in the structure.

vi. **Continually update the list of structural elements** or fragments that have been conclusively identified at each step and start to pull together reasonable possible structures. Be careful not to jump to possible solutions before the evidence is conclusive. Keep assessing and re-assessing all of the options.

vii. When you have a final solution which you believe is correct, **go back and confirm that all of the spectroscopic data are consistent with the final structure**.

5.2 WORKED EXAMPLE 1

Identify the following compound.

Molecular Formula: $C_5H_{10}O$

IR: 1721 cm^{-1}

^{1}H NMR Spectrum
(CDCl$_3$, 500 MHz)

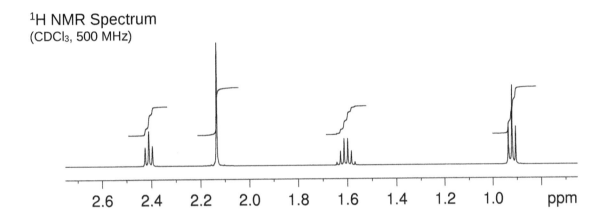

^{13}C{^{1}H} NMR Spectrum
(CDCl$_3$, 125 MHz)

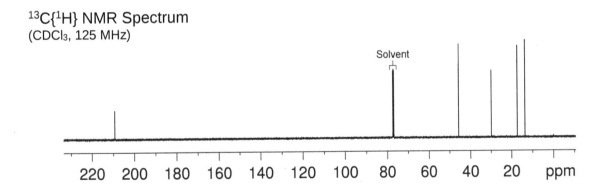

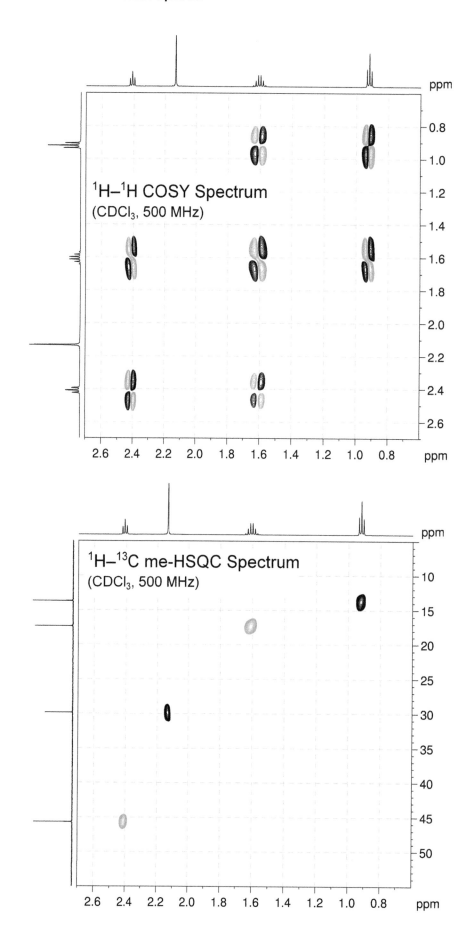

¹H–¹H COSY Spectrum
(CDCl₃, 500 MHz)

¹H–¹³C me-HSQC Spectrum
(CDCl₃, 500 MHz)

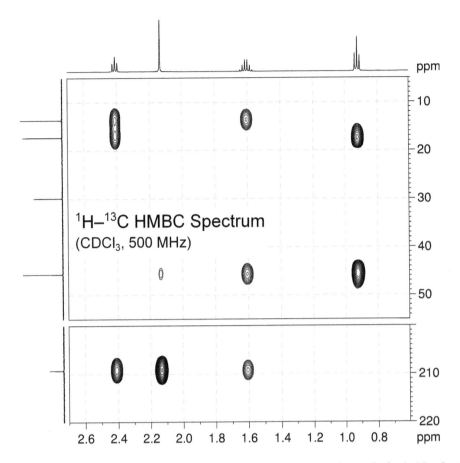

^{1}H–^{13}C HMBC Spectrum
(CDCl$_3$, 500 MHz)

i.   ***Determine the degree of unsaturation.*** The molecular formula is C$_5$H$_{10}$O. Ignore the O atom to give an effective molecular formula of C$_5$H$_{10}$ (C$_n$H$_m$). The degree of unsaturation is given by

$$\text{Degree of unsaturation} = n - \frac{m}{2} + 1$$
$$= 5 - \frac{10}{2} + 1$$
$$= 1$$

The compound must contain the equivalent of one π bond or ring.

ii. ***Determine the number of protons in each unique environment.*** Measure the total integral across all signals in the ^{1}H spectrum. From the molecular formula, there are 10 protons in the structure, so work out the integral that corresponds to one proton and then the relative numbers of protons in different environments.

59

$\delta\ ^1H$ (ppm)	Integral (mm)	Relative Number of Hydrogens	Multiplicity
0.92	12	3	Triplet
1.60	8	2	Sextet
2.13	12	3	Singlet
2.41	8	2	Triplet

Note that this analysis gives a total of 10 protons, which is consistent with the molecular formula. The multiplicities of the signals can be verified using the me-HSQC spectrum.

iii. **Determine the number of unique carbon environments.** From the $^{13}C\{^1H\}$ NMR spectrum, there are five unique carbon environments. One carbon is in the typical carbonyl chemical shift range, five are in the aliphatic chemical shift range. From the me-HSQC spectrum, there are two $-CH_2-$ groups and two resonances from either $-CH-$ or $-CH_3$ groups. The $^{13}C\{^1H\}$ spectrum establishes that the compound contains a ketone (^{13}C resonance at 209.4 ppm). There can be no other double bonds or rings in the molecule because the C=O accounts for the single degree of unsaturation.

iv. **Identify any structural elements.**

   a. The $^{13}C\{^1H\}$ spectrum establishes that the compound contains a ketone (^{13}C resonance at 209.4 ppm).

   b. The COSY spectrum shows two distinct spin systems – one containing a $CH_3CH_2CH_2$ fragment, the other a single CH_3 group.

   c. The $^1H-^{13}C$ me-HSQC spectrum easily identifies the protonated carbon resonances at 13.7, 17.3, 29.9 and 45.7 ppm.

v. **Identify possible candidates.** Only one possible isomer may be constructed using these structural elements:

$$\overset{\displaystyle O}{\underset{}{\parallel}}$$
$$\underset{1}{CH_3}-\underset{2}{C}-\underset{3}{CH_2}-\underset{4}{CH_2}-\underset{5}{CH_3}$$

vi. **Use the 2D spectra to confirm the structure of the compound.** The HMBC spectrum confirms the structure with correlations from H_1 and H_3 to C_2, indicating that the ketone group is located between C_1 and C_3.

vii. ***Re-inspect all spectra.*** Re-inspect all the spectra to ensure that every peak is consistent with the proposed structure.

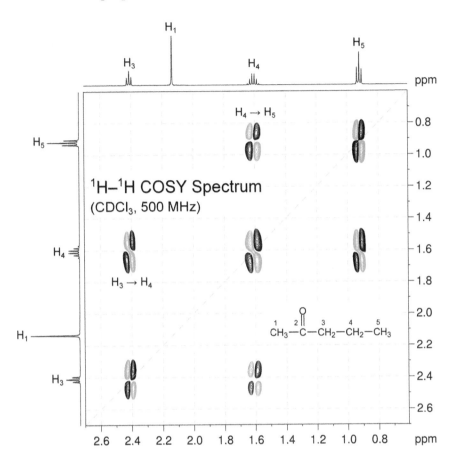

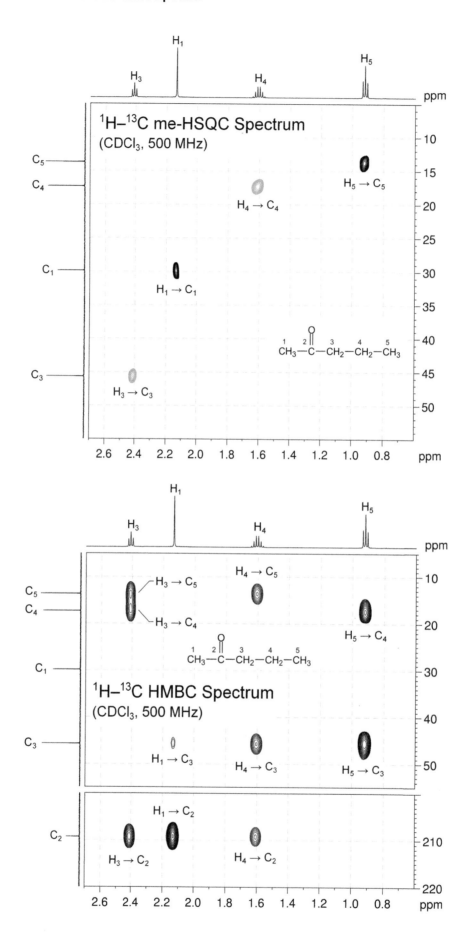

5.3 WORKED EXAMPLE 2

Identify the following compound.

Molecular Formula: $C_{11}H_{14}O$

IR: 1683 cm^{-1}

^{1}H NMR Spectrum
(CDCl$_3$, 400 MHz)

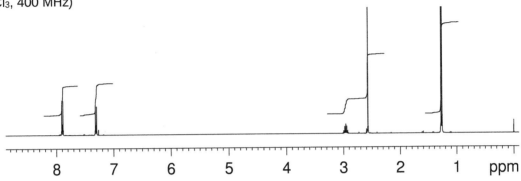

^{1}H NMR Expansions
(CDCl$_3$, 400 MHz)

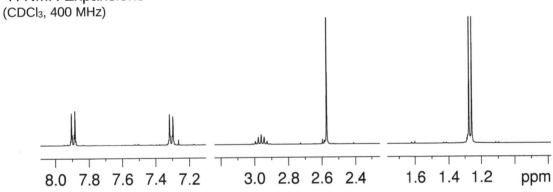

^{13}C{^{1}H} NMR Spectrum
(CDCl$_3$, 100 MHz)

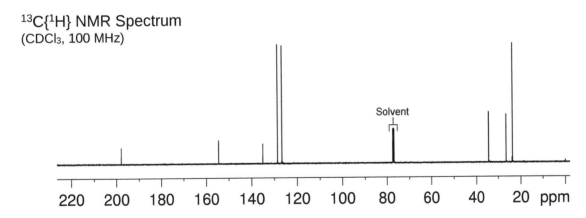

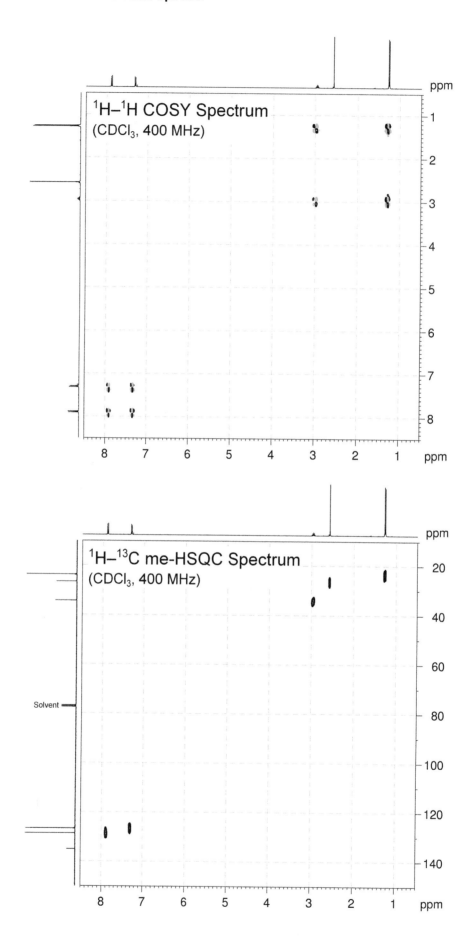

¹H–¹H COSY Spectrum
(CDCl₃, 400 MHz)

¹H–¹³C me-HSQC Spectrum
(CDCl₃, 400 MHz)

Solvent

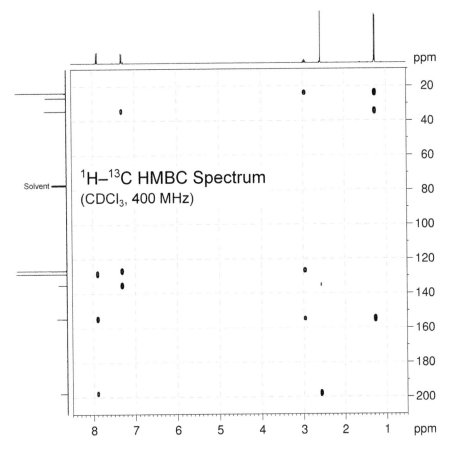

i. ***Determine the degree of unsaturation.*** The molecular formula is $C_{11}H_{14}O$. Omit the oxygen to give an effective molecular formula of $C_{11}H_{14}$ (C_nH_m). The degree of unsaturation is given by

$$
\begin{aligned}
\text{Degree of unsaturation} \quad &= \quad n - \frac{m}{2} + 1 \\
&= \quad 11 - \frac{14}{2} + 1 \\
&= \quad 5
\end{aligned}
$$

The compound must contain a combined total of five π bonds and/or rings.

The ^{1}H and ^{13}C spectra contain resonances in the aromatic region of the spectra. An aromatic ring would account for four degrees of unsaturation so the molecule must contain an aromatic ring and one other π bond or ring.

ii. ***Determine the number of protons in each unique environment.*** Measure the total integral across all signals in the ^{1}H spectrum. From the molecular formula, there are 14 protons in the structure, so evaluate the integral that corresponds to one proton and then establish the relative numbers of protons in different environments.

$\delta\,^1H$ (ppm)	Integral (mm)	Relative Number of Hydrogens (rounded)	Multiplicity
1.28	23	6	Doublet
2.59	11	3	Singlet
2.96	4	1	Septet
7.31	8	2	Multiplet
7.89	8	2	Multiplet

Note that this analysis gives a total of 14 protons, which is consistent with the molecular formula.

iii. ***Determine the number of unique carbon environments.*** From the $^{13}C\{^1H\}$ NMR spectrum, there are eight carbon environments: one carbon is in the typical carbonyl chemical shift range, four carbons are in the aromatic chemical shift range and three are in the aliphatic chemical shift range. From the me-HSQC spectrum, there are no $-CH_2-$ groups – all of the carbons are either $-CH-$ or $-CH_3$ groups. There are 11 carbons in the molecular formula, and only eight unique environments, so the compound must contain some symmetry elements.

iv. ***Identify any structural elements.***

    a. The coupling pattern in the aromatic region of the ^1H NMR spectrum is consistent with a 1,4-disubstituted benzene, in which the two ring substituents are not identical. The number of signals in the aromatic region of the $^{13}C\{^1H\}$ NMR spectrum is also consistent with this substitution pattern.

    b. In the aliphatic region of the ^1H NMR spectrum, the septet coupling of the one-proton signal at 2.96 ppm is consistent with this proton being adjacent to six neighbouring protons. Similarly, the doublet coupling of the six-proton signal at 1.28 ppm is consistent with these protons having a single neighbour. An isolated *iso*-propyl group ($-CH(CH_3)_2$) is therefore present and also consistent with the cross-peak observed in the COSY spectrum.

    c. The remaining signal in the ^1H NMR spectrum (a singlet of integration 3H) is consistent with the presence of an isolated methyl ($-CH_3$) group.

    d. The signal at 197 ppm in the $^{13}C\{^1H\}$ NMR spectrum suggests the presence of a ketone functional group. This is confirmed by the IR absorption at

1683 cm^{-1}. The benzene ring and the ketone group account for all the degrees of unsaturation in the molecule.

e. The structural elements are therefore:

1. ![1,4-disubstituted benzene ring structure]

2. $-CH(CH_3)_2$

3. $-CH_3$

4. $-C(=O)-$

v. **_Identify possible candidates._** Two possible isomers may be constructed using these structural elements:

![Structure A: CH3-benzene-C(=O)-CH(CH3)2; Structure B: (CH3)2CH-benzene-C(=O)-CH3]

A   **B**

Both isomers are entirely consistent with the available one-dimensional NMR data.

vi. **_Use the 2D spectra to identify the correct compound._**

a. The ^1H–^1H COSY spectrum confirms the presence of coupling between the isopropyl CH and CH_3 groups, while the ^1H–^{13}C me-HSQC allows assignment of all the protonated carbon atoms in the structure.

b. The ^1H–^{13}C HMBC allows us to differentiate between the two possible structures. There is a strong cross-peak between the protons of the isolated methyl group and the carbonyl carbon. In structure A, the methyl protons are six bonds away from the carbonyl carbon; this is too far to experience any coupling (HMBC correlations are typically limited to couplings over two- or three-bond separations). In structure B, the methyl protons are two bonds away from the carbonyl carbon, so we can confidently conclude that **Structure B** (4'-isopropylacetophenone) is the correct isomer.

c. If required, the HMBC can also be used to assign all the non-protonated carbon atoms in the molecule. In the HMBC spectrum, there is a correlation between H_1 and $C_{1'}$. While this

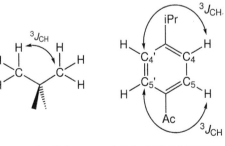

gem-dimethyl   1,4-disubstituted benzene

appears to be a one-bond correlation, in *gem*-dimethyl groups, the apparent one-bond correlation arises from the $^3J_{CH}$ interaction of the protons of one of the methyl groups with the chemically equivalent carbon which is three bonds away. Similarly, there are correlations between H_4 and $C_{4'}$, and H_5 and $C_{5'}$. These arise from three-bond correlations between the proton and the symmetry-related *meta* carbon.

vii. **Re-inspect all spectra.** Re-inspect all the spectra to ensure that every peak is consistent with the proposed structure.

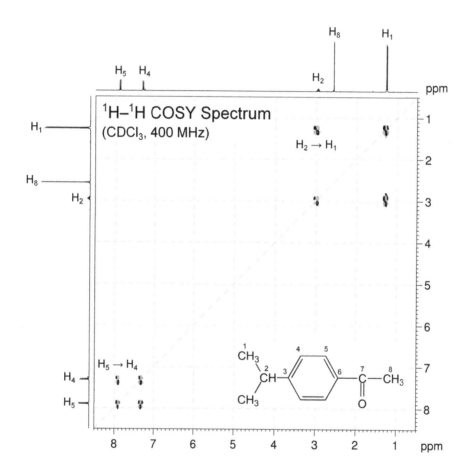

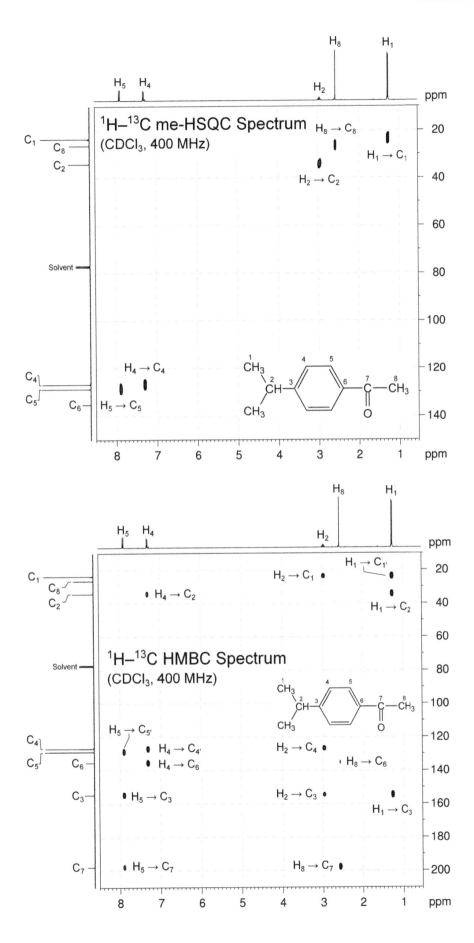

6 Problems

Problem 1

The ^{1}H and ^{13}C{^{1}H} NMR spectra of 1-iodopropane (C$_3$H$_7$I) recorded in CDCl$_3$ solution at 298 K and 400 MHz are given below.

The ^{1}H NMR spectrum has signals at δ 0.99 (H$_3$), 1.84 (H$_2$) and 3.18 (H$_1$) ppm.

The ^{13}C{^{1}H} NMR spectrum has signals at δ 9.6 (C$_1$), 15.3 (C$_3$) and 26.9 (C$_2$) ppm.

Also given on the following pages are the ^{1}H–^{1}H COSY, ^{1}H–^{13}C me-HSQC, ^{1}H–^{13}C HMBC and INADEQUATE spectra. For each 2D spectrum, indicate which correlation gives rise to each cross-peak by placing an appropriate label in the box provided (e.g. H$_1 \rightarrow$ H$_2$, H$_1 \rightarrow$ C$_1$).

$$\begin{array}{ccc} 1 & 2 & 3 \end{array}$$
$$\text{I–CH}_2\text{–CH}_2\text{–CH}_3$$

^{1}H NMR Spectrum
(CDCl$_3$, 400 MHz)

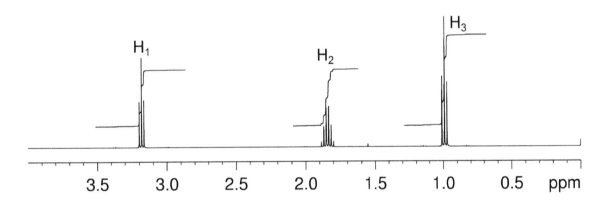

^{13}C{^{1}H} NMR Spectrum
(CDCl$_3$, 100 MHz)

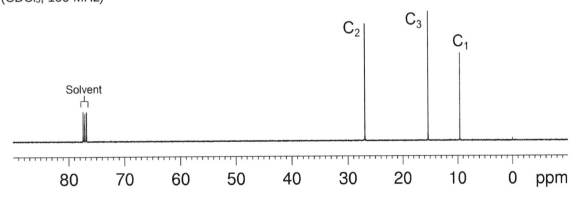

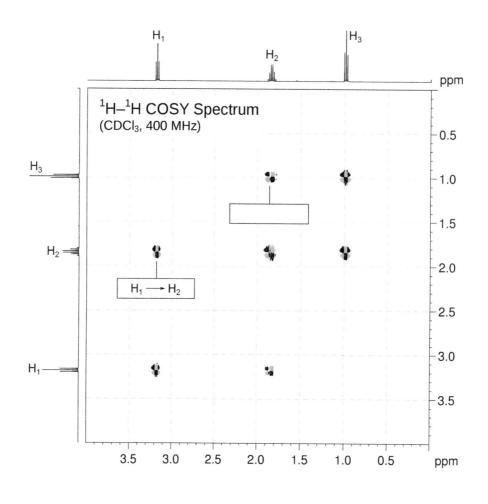

¹H–¹H COSY Spectrum
(CDCl₃, 400 MHz)

H₁ ⟶ H₂

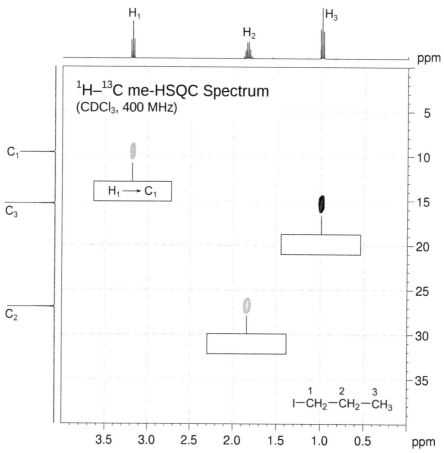

¹H–¹³C me-HSQC Spectrum
(CDCl₃, 400 MHz)

H₁ ⟶ C₁

$$I-\underset{1}{CH_2}-\underset{2}{CH_2}-\underset{3}{CH_3}$$

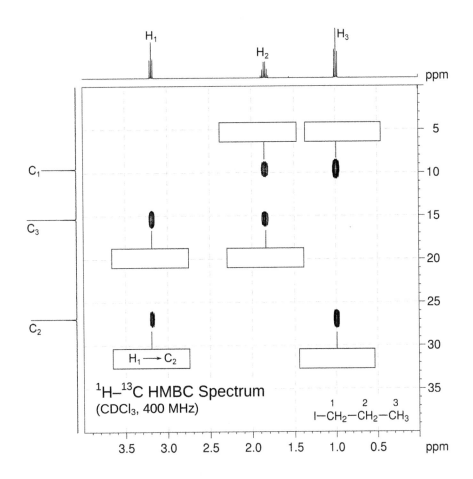

^{1}H–^{13}C HMBC Spectrum
(CDCl$_3$, 400 MHz)

H$_1$ → C$_2$

$\underset{1}{I-CH_2}-\underset{2}{CH_2}-\underset{3}{CH_3}$

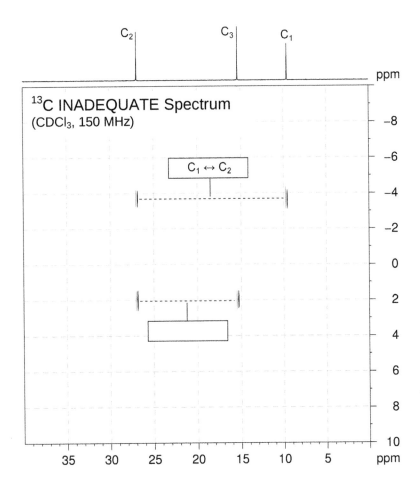

^{13}C INADEQUATE Spectrum
(CDCl$_3$, 150 MHz)

C$_1$ ↔ C$_2$

Problem 2

The ^{1}H and $^{13}C\{^{1}H\}$ NMR spectra of 2-butanone (C_4H_8O) recorded in $CDCl_3$ solution at 298 K and 400 MHz are given below.

The ^{1}H NMR spectrum has signals at δ 1.05 (H_4), 2.14 (H_1) and 2.47 (H_3) ppm.

The $^{13}C\{^{1}H\}$ NMR spectrum has signals at δ 7.2 (C_4), 28.8 (C_1), 36.2 (C_3) and 208.8 (C_2) ppm.

Also given on the following pages are the ^{1}H–^{1}H COSY, ^{1}H–^{13}C me-HSQC, ^{1}H–^{13}C HMBC and INADEQUATE spectra. For each 2D spectrum, indicate which correlation gives rise to each cross-peak by placing an appropriate label in the box provided (*e.g.* $H_1 \rightarrow H_2$, $H_1 \rightarrow C_1$).

$$\underset{O}{\overset{\displaystyle 1 \quad\; 2 \quad\;\; 3 \quad\;\; 4}{CH_3-C-CH_2-CH_3}}$$

^{1}H NMR Spectrum
(CDCl$_3$, 400 MHz)

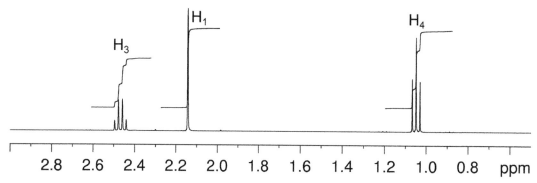

$^{13}C\{^{1}H\}$ NMR Spectrum
(CDCl$_3$, 100 MHz)

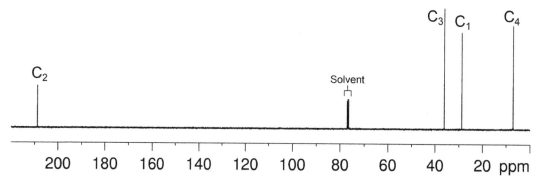

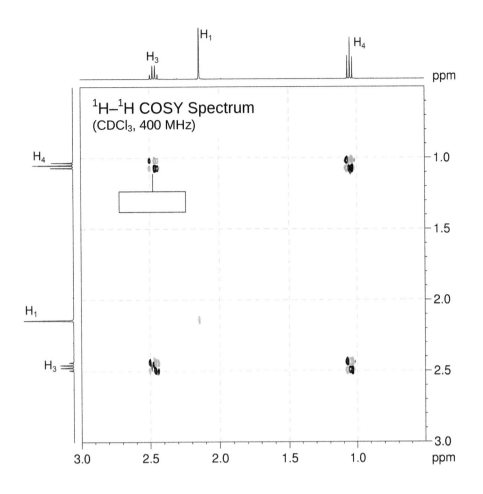

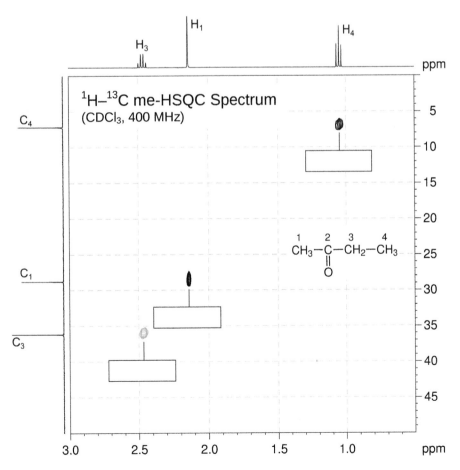

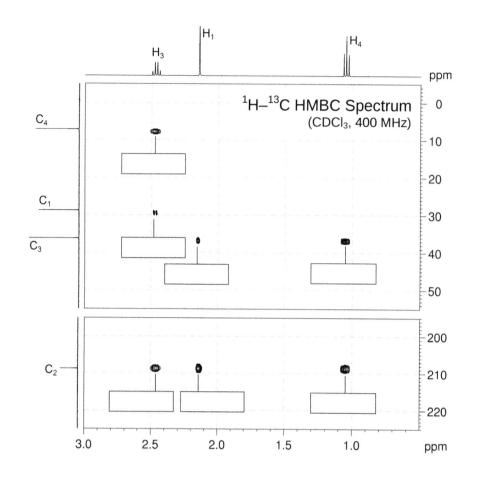

¹H–¹³C HMBC Spectrum
(CDCl$_3$, 400 MHz)

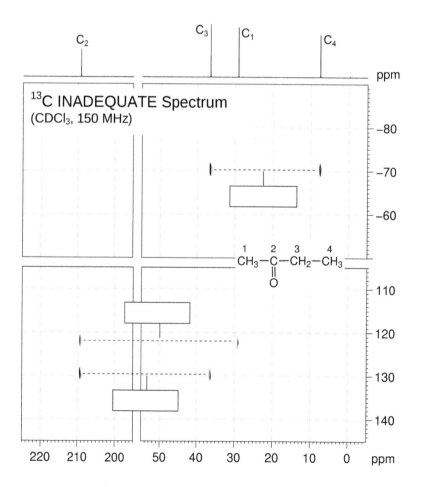

¹³C INADEQUATE Spectrum
(CDCl$_3$, 150 MHz)

$$\underset{\substack{\|\\O}}{\overset{\displaystyle 1\quad\quad 2\quad\quad 3\quad\quad 4}{CH_3-C-CH_2-CH_3}}$$

Problem 3

Identify the following compound.
Molecular Formula: $C_6H_{12}O$
IR: 1718 cm^{-1}

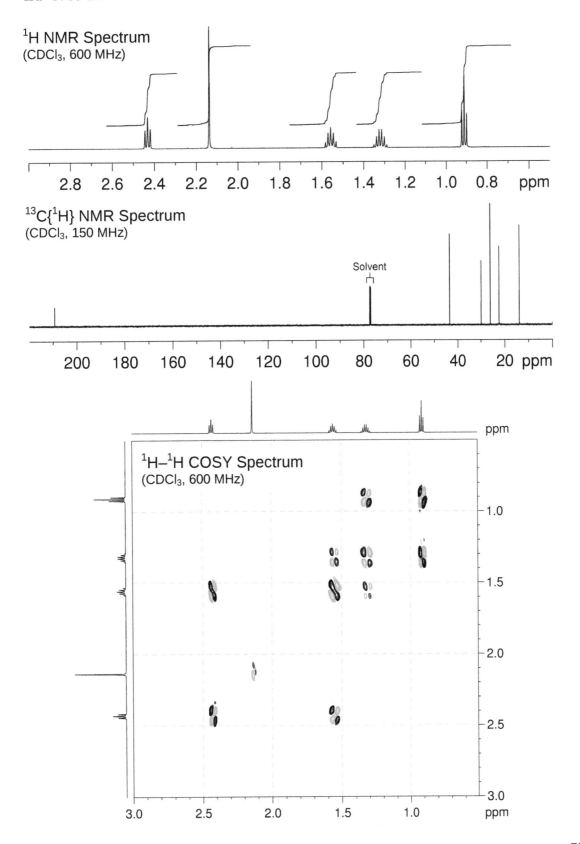

^{1}H NMR Spectrum
(CDCl$_3$, 600 MHz)

^{13}C{^{1}H} NMR Spectrum
(CDCl$_3$, 150 MHz)

Solvent

^{1}H–^{1}H COSY Spectrum
(CDCl$_3$, 600 MHz)

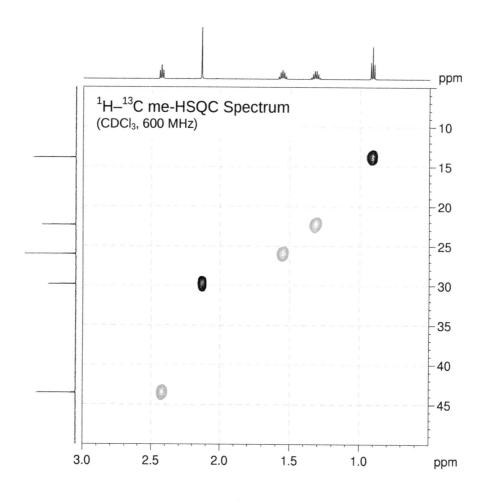

¹H–¹³C me-HSQC Spectrum
(CDCl₃, 600 MHz)

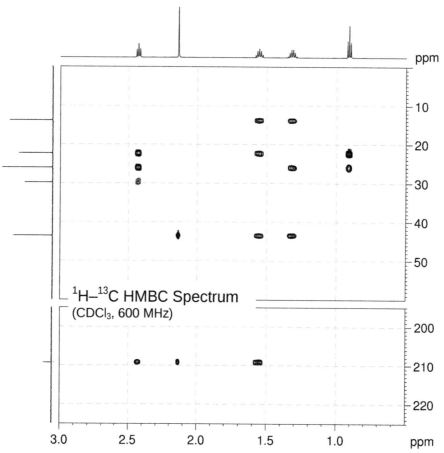

¹H–¹³C HMBC Spectrum
(CDCl₃, 600 MHz)

Problem 4

The ^{1}H and ^{13}C$\{^1$H$\}$ NMR spectra of ethyl propionate ($C_5H_{10}O_2$) recorded in $CDCl_3$ solution at 298 K and 300 MHz are given below.

The ^{1}H NMR spectrum has signals at δ 1.14 (H_1), 1.26 (H_5), 2.31 (H_2) and 4.12 (H_4) ppm.

The ^{13}C$\{^1$H$\}$ NMR spectrum has signals at δ 9.2 (C_1), 14.3 (C_5), 27.7 (C_2), 60.3 (C_4) and 174.5 (C_3) ppm.

Use this information to produce schematic diagrams of the COSY, HSQC and HMBC spectra, showing where all of the cross peaks and diagonal peaks would be.

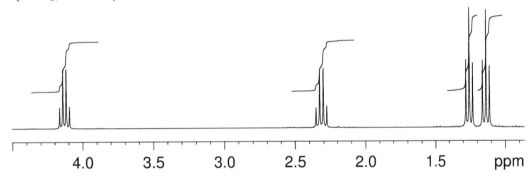

^{1}H NMR Spectrum
(CDCl$_3$, 300 MHz)

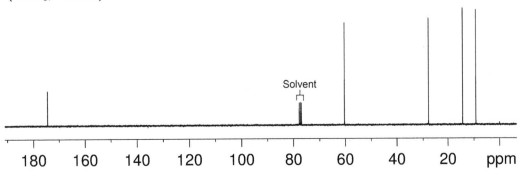

^{13}C$\{^1$H$\}$ NMR Spectrum
(CDCl$_3$, 75 MHz)

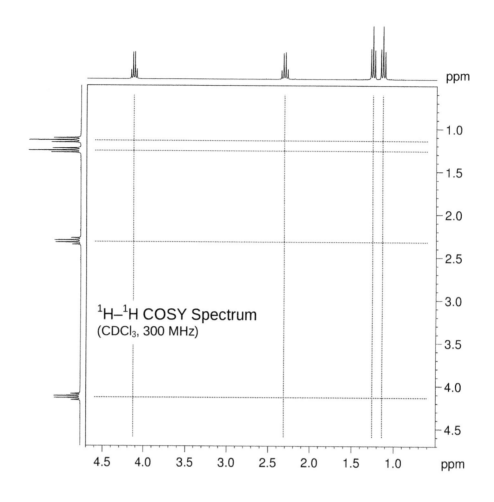

^{1}H–^{1}H COSY Spectrum
(CDCl$_3$, 300 MHz)

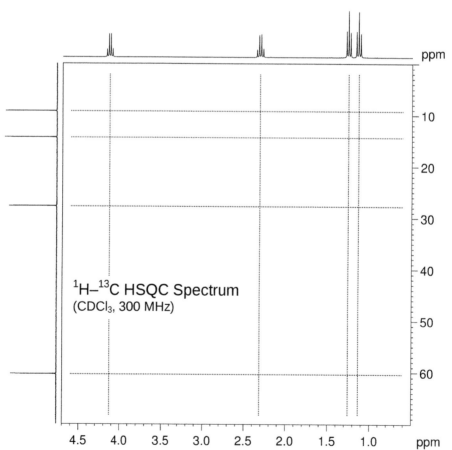

^{1}H–^{13}C HSQC Spectrum
(CDCl$_3$, 300 MHz)

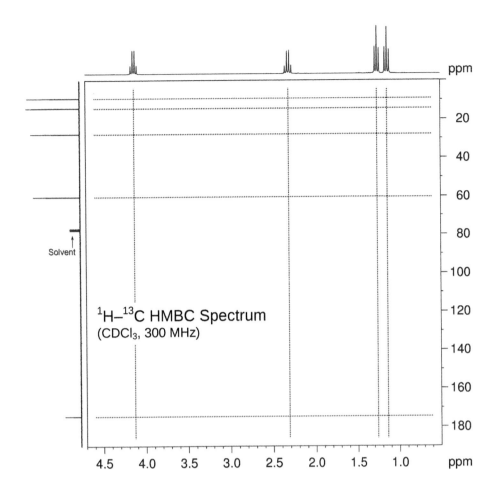

^{1}H–^{13}C HMBC Spectrum
(CDCl$_3$, 300 MHz)

Solvent

4.5	4.0	3.5	3.0	2.5	2.0	1.5	1.0	ppm

ppm — 20, 40, 60, 80, 100, 120, 140, 160, 180

$$\underset{1}{CH_3}-\underset{2}{CH_2}-\underset{3}{\underset{\displaystyle \|}{\underset{O}{C}}}-O-\underset{4}{CH_2}-\underset{5}{CH_3}$$

Problem 5

The ^{1}H and ^{13}C{^{1}H} NMR spectra of ethyl 3-ethoxypropionate ($C_7H_{14}O_3$) recorded in CDCl$_3$ solution at 298 K and 600 MHz are given below.

The ^{1}H NMR spectrum has signals at δ 1.18 (H$_1$), 1.26 (H$_7$), 2.56 (H$_4$), 3.50 (H$_2$), 3.70 (H$_3$) and 4.15 (H$_6$) ppm.

The ^{13}C{^{1}H} NMR spectrum has signals at δ 14.2 (C$_7$), 15.1 (C$_1$), 35.3 (C$_4$), 60.4 (C$_6$), 65.9 (C$_3$), 66.4 (C$_2$) and 171.7 (C$_5$) ppm.

Also given on the following pages are the ^{1}H–^{1}H COSY, ^{1}H–^{13}C me-HSQC and ^{1}H–^{13}C HMBC spectra. For each 2D spectrum, indicate which correlation gives rise to each cross-peak by placing an appropriate label in the box provided (*e.g.* H$_1$ → H$_2$, H$_1$ → C$_1$).

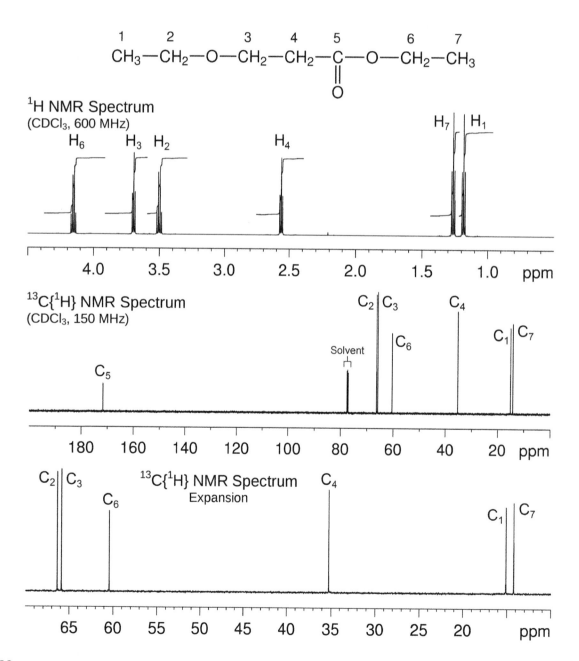

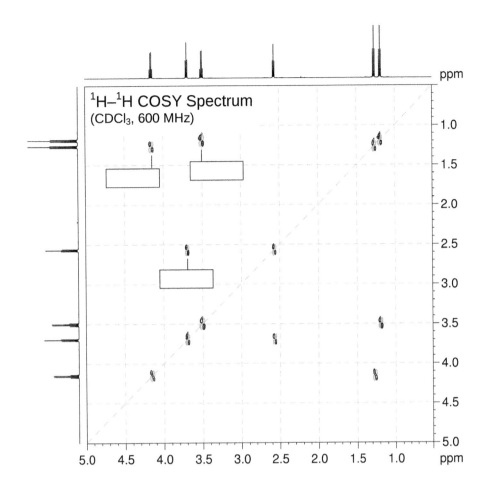

¹H–¹H COSY Spectrum
(CDCl₃, 600 MHz)

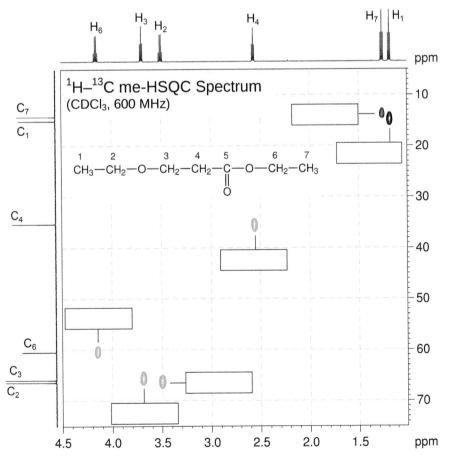

¹H–¹³C me-HSQC Spectrum
(CDCl₃, 600 MHz)

$$CH_3-CH_2-O-CH_2-CH_2-C-O-CH_2-CH_3$$

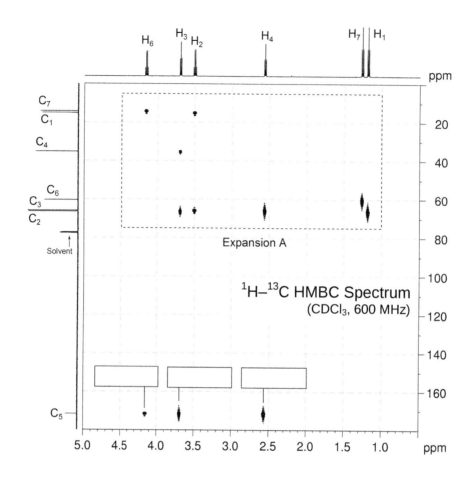

Expansion A

^1H–^{13}C HMBC Spectrum
(CDCl$_3$, 600 MHz)

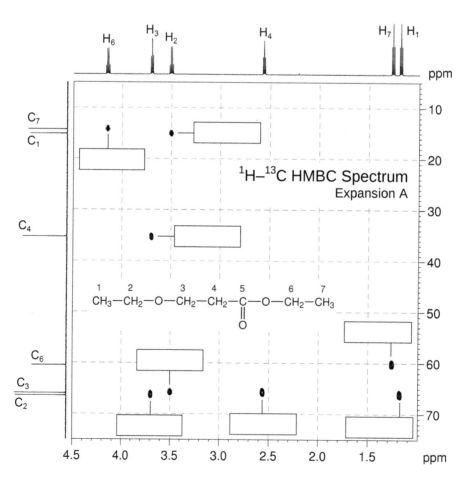

^1H–^{13}C HMBC Spectrum
Expansion A

$$\underset{1}{CH_3}-\underset{2}{CH_2}-O-\underset{3}{CH_2}-\underset{4}{CH_2}-\underset{5}{\underset{\underset{O}{\|}}{C}}-O-\underset{6}{CH_2}-\underset{7}{CH_3}$$

Problem 6

The ^1H and $^{13}C\{^1H\}$ NMR spectra of 4-acetylbutyric acid ($C_6H_{10}O_3$) recorded in CDCl$_3$ solution at 298 K and 600 MHz are given below. The ^1H NMR spectrum has signals at δ 1.81, 2.08, 2.31, 2.47 and 10.5 ppm. The $^{13}C\{^1H\}$ NMR spectrum has signals at δ 18.5, 29.8, 32.9, 42.2, 178.8 and 208.6 ppm. The ^1H–^{13}C me-HSQC and ^1H–^{13}C HMBC spectra are given on the following pages. Use these spectra to assign the ^1H and $^{13}C\{^1H\}$ resonances for this compound.

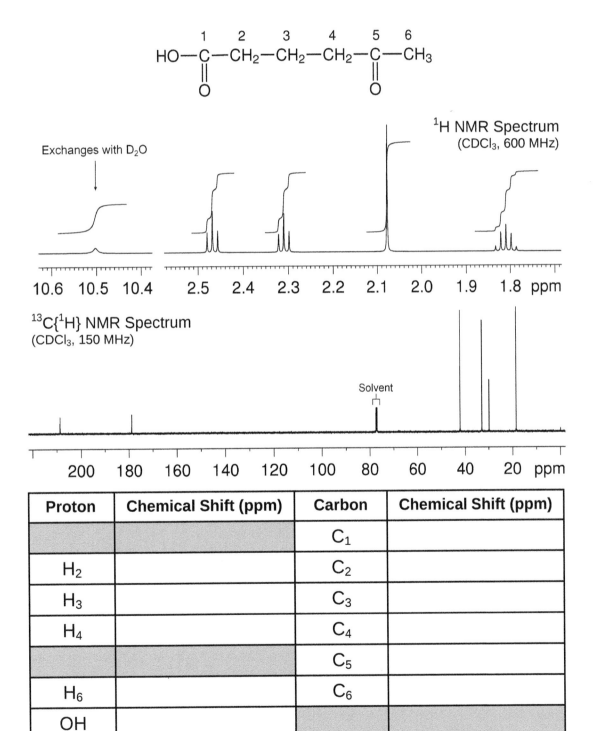

Proton	Chemical Shift (ppm)	Carbon	Chemical Shift (ppm)
		C$_1$	
H$_2$		C$_2$	
H$_3$		C$_3$	
H$_4$		C$_4$	
		C$_5$	
H$_6$		C$_6$	
OH			

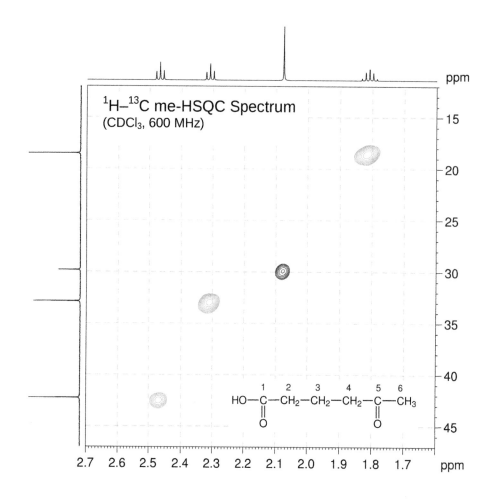

$^1H–^{13}C$ me-HSQC Spectrum
(CDCl$_3$, 600 MHz)

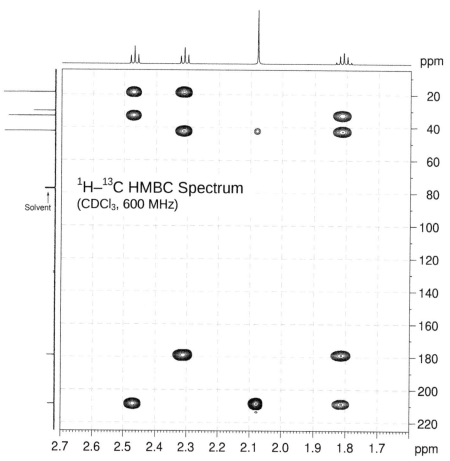

$^1H–^{13}C$ HMBC Spectrum
(CDCl$_3$, 600 MHz)

Solvent

Problem 7

Identify the following compound.
Molecular Formula: $C_5H_9ClO_2$

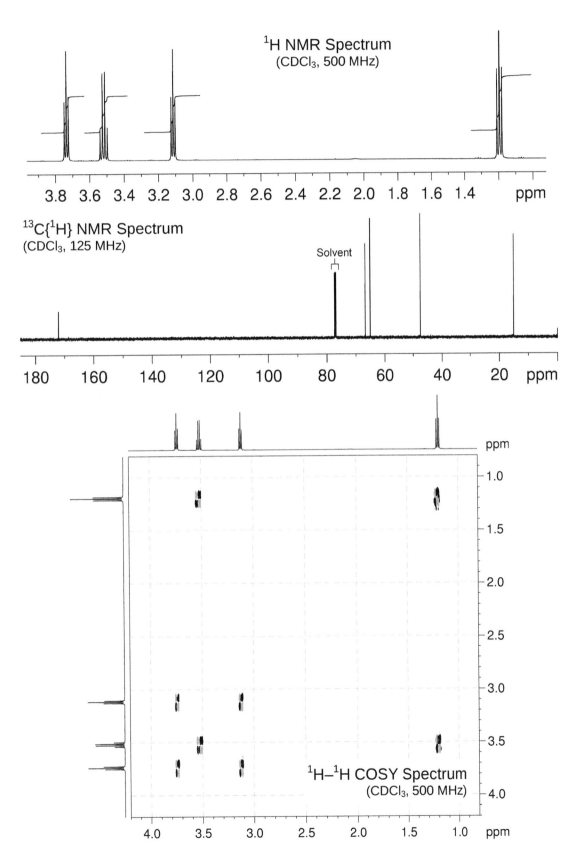

^{1}H NMR Spectrum
(CDCl$_3$, 500 MHz)

^{13}C{^{1}H} NMR Spectrum
(CDCl$_3$, 125 MHz)

Solvent

^{1}H–^{1}H COSY Spectrum
(CDCl$_3$, 500 MHz)

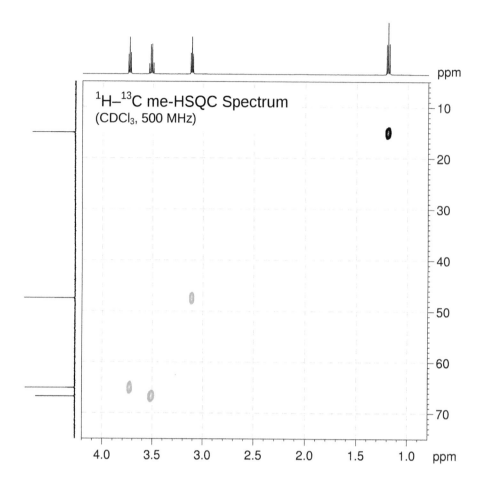

^{1}H–^{13}C me-HSQC Spectrum
(CDCl$_3$, 500 MHz)

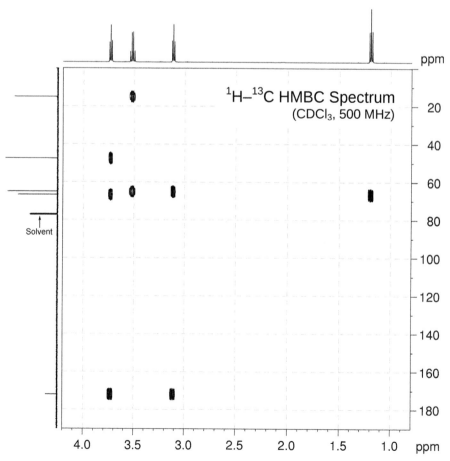

^{1}H–^{13}C HMBC Spectrum
(CDCl$_3$, 500 MHz)

Solvent

Problem 8

Identify the following compound.
Molecular Formula: $C_5H_9ClO_2$

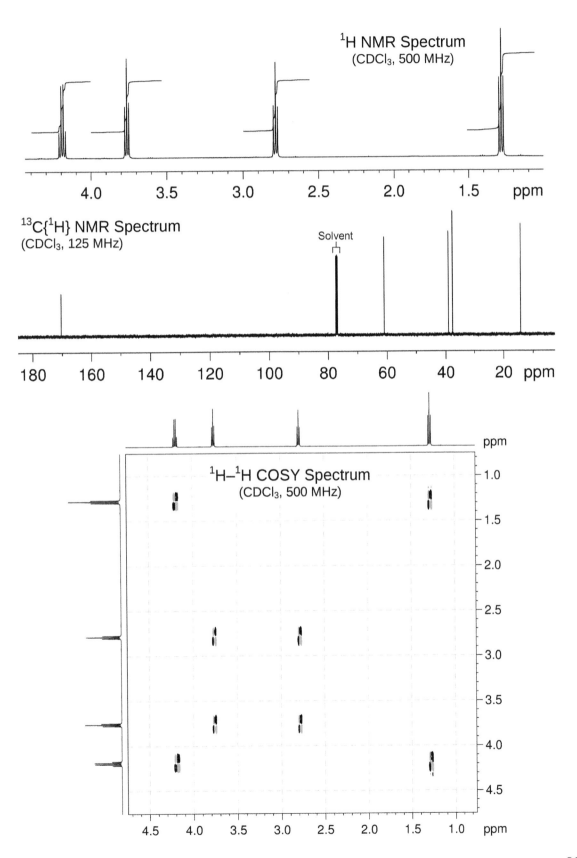

^{1}H NMR Spectrum
(CDCl$_3$, 500 MHz)

^{13}C{^{1}H} NMR Spectrum
(CDCl$_3$, 125 MHz)

Solvent

^{1}H–^{1}H COSY Spectrum
(CDCl$_3$, 500 MHz)

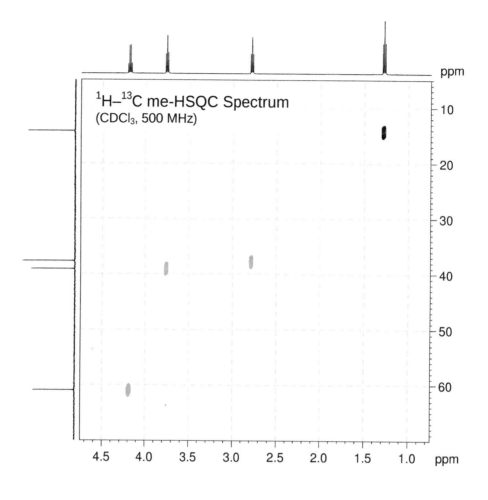

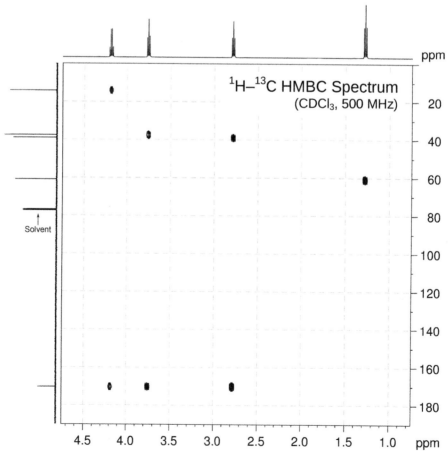

Problem 9

The ^{1}H and ^{13}C{^{1}H} NMR spectra of isoamyl acetate ($C_7H_{14}O_2$) recorded in CDCl$_3$ solution at 298 K and 600 MHz are given below.

The ^{1}H NMR spectrum has signals at δ 0.92 (H$_6$), 1.52 (H$_4$), 1.69 (H$_5$), 2.04 (H$_1$) and 4.09 (H$_3$) ppm.

The ^{13}C{^{1}H} NMR spectrum has signals at δ 21.0 (C$_1$), 22.5 (C$_6$), 25.1 (C$_5$), 37.4 (C$_4$), 63.1 (C$_3$) and 171.2 (C$_2$) ppm.

Also given on the following pages are the ^{1}H–^{1}H COSY, ^{1}H–^{13}C me-HSQC, ^{1}H–^{13}C HMBC and INADEQUATE spectra. For each 2D spectrum, indicate which correlation gives rise to each cross-peak by placing an appropriate label in the box provided.

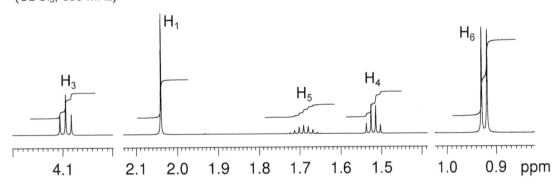

^{1}H NMR Spectrum
(CDCl$_3$, 600 MHz)

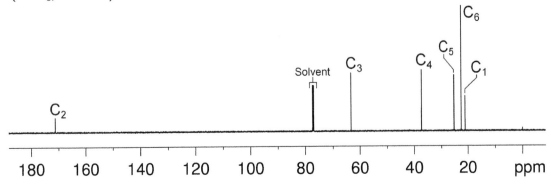

^{13}C{^{1}H} NMR Spectrum
(CDCl$_3$, 150 MHz)

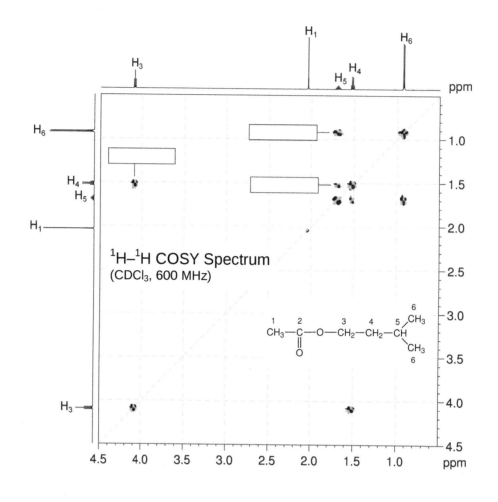

^1H–^1H COSY Spectrum
(CDCl$_3$, 600 MHz)

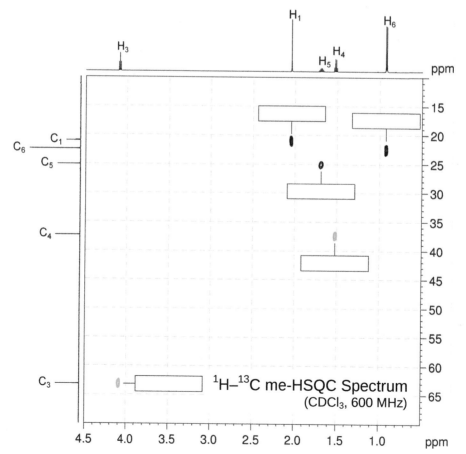

^1H–^{13}C me-HSQC Spectrum
(CDCl$_3$, 600 MHz)

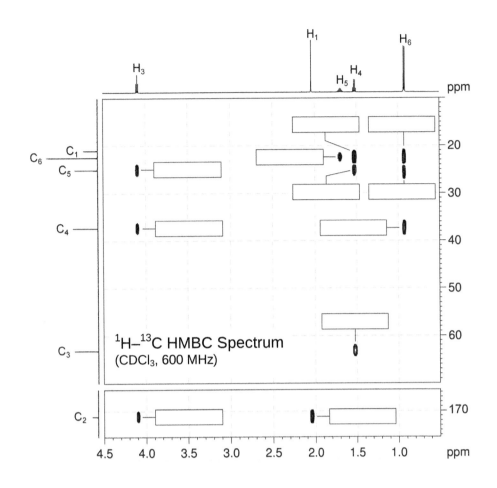

^{1}H–^{13}C HMBC Spectrum
(CDCl$_3$, 600 MHz)

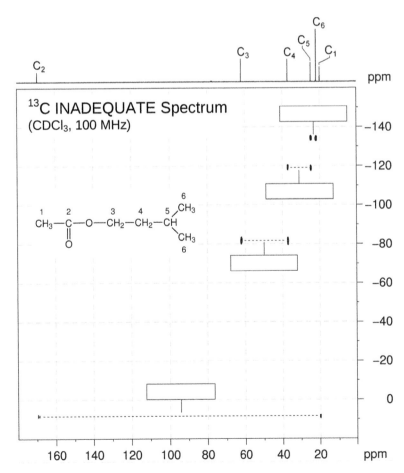

^{13}C INADEQUATE Spectrum
(CDCl$_3$, 100 MHz)

93

Problem 10

The ^1H and $^{13}C\{^1H\}$ NMR spectra of *trans*-4-hexen-3-one ($C_6H_{10}O$) recorded in DMSO-d_6 solution at 298 K and 400 MHz are given below.

The ^1H NMR spectrum has signals at δ 0.96 (H_1), 1.86 (H_6), 2.56 (H_2), 6.11 (H_4) and 6.85 (H_5) ppm.

The $^{13}C\{^1H\}$ NMR spectrum has signals at δ 8.4 (C_1), 18.4 (C_6), 32.6 (C_2), 131.9 (C_4), 142.8 (C_5) and 200.4 (C_3) ppm.

Also given on the following pages are the ^1H–^1H COSY, ^1H–^{13}C me-HSQC, ^1H–^{13}C HMBC and ^1H–^1H NOESY spectra. For each 2D spectrum, indicate which correlation gives rise to each cross-peak by placing an appropriate label in the box provided.

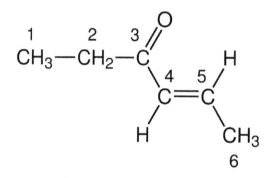

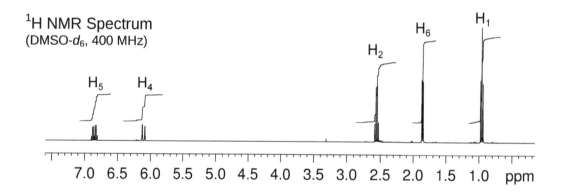

^1H NMR Spectrum
(DMSO-d_6, 400 MHz)

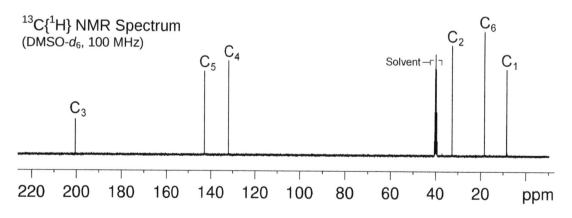

$^{13}C\{^1H\}$ NMR Spectrum
(DMSO-d_6, 100 MHz)

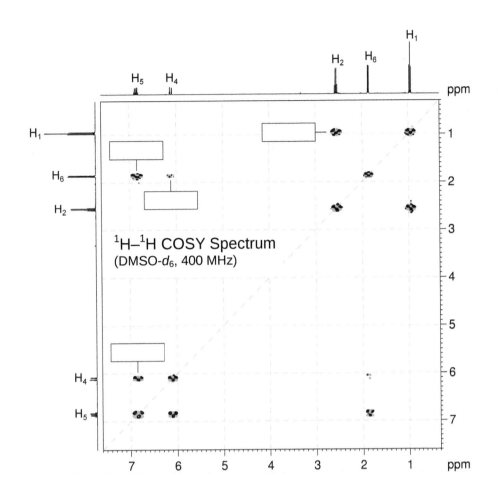

^{1}H–^{1}H COSY Spectrum
(DMSO-d_6, 400 MHz)

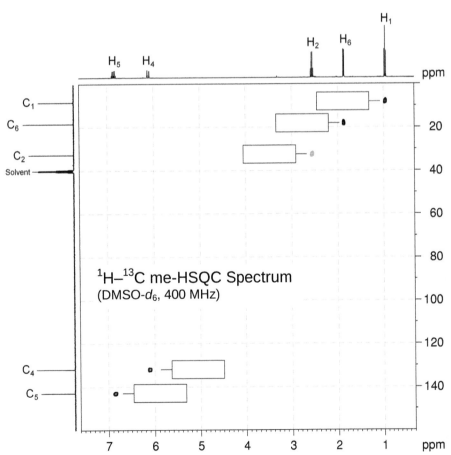

^{1}H–^{13}C me-HSQC Spectrum
(DMSO-d_6, 400 MHz)

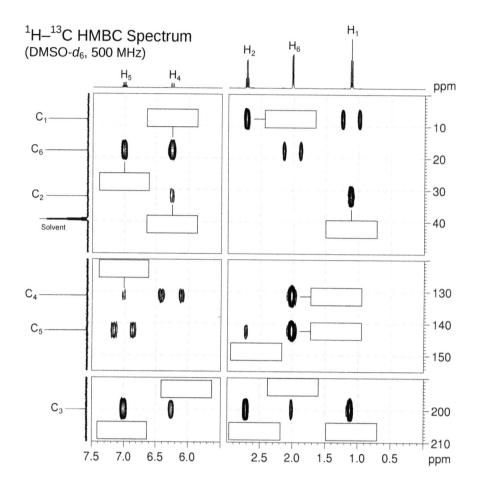

¹H–¹³C HMBC Spectrum
(DMSO-d_6, 500 MHz)

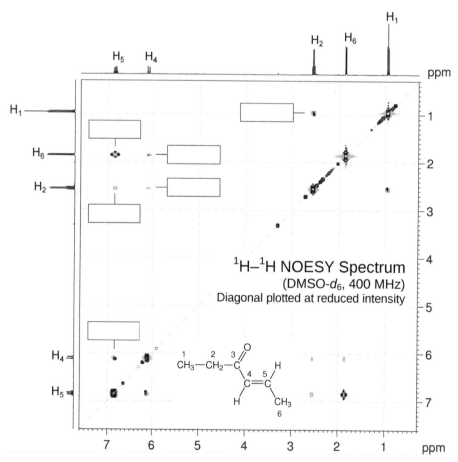

¹H–¹H NOESY Spectrum
(DMSO-d_6, 400 MHz)
Diagonal plotted at reduced intensity

Problem 11

Identify the following compound.
Molecular Formula: $C_8H_{14}O$
IR: 1698, 1638 cm^{-1}

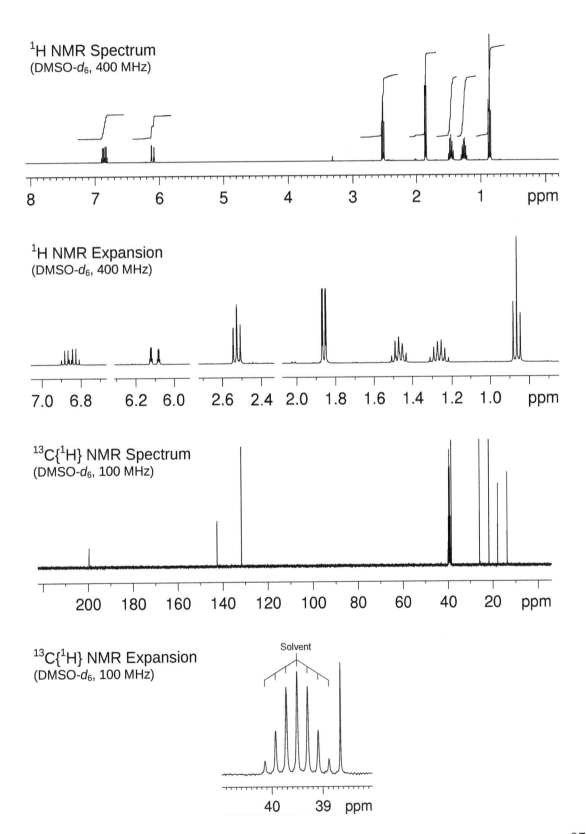

^{1}H NMR Spectrum
(DMSO-d_6, 400 MHz)

^{1}H NMR Expansion
(DMSO-d_6, 400 MHz)

^{13}C{^{1}H} NMR Spectrum
(DMSO-d_6, 100 MHz)

^{13}C{^{1}H} NMR Expansion
(DMSO-d_6, 100 MHz)

Solvent

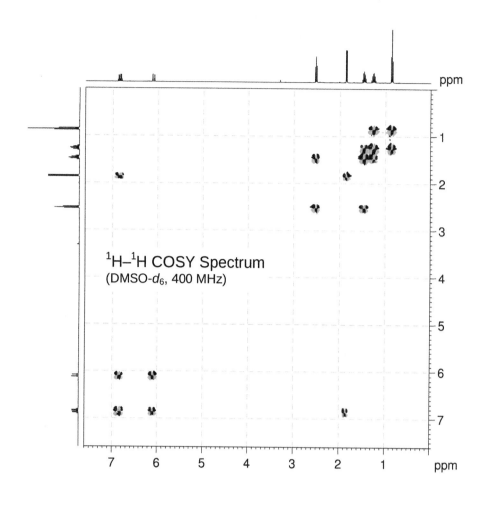

^1H–^1H COSY Spectrum
(DMSO-d_6, 400 MHz)

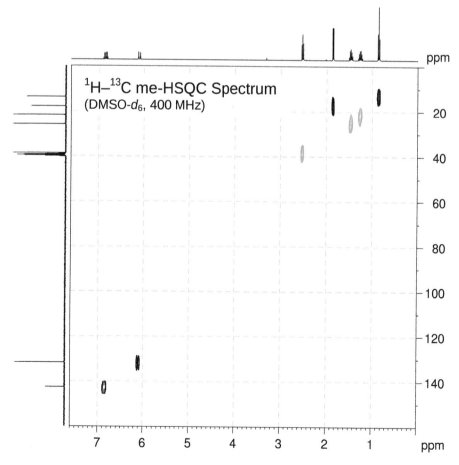

^1H–^{13}C me-HSQC Spectrum
(DMSO-d_6, 400 MHz)

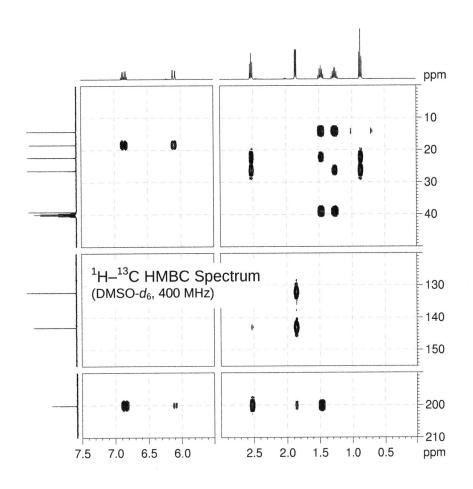

$^1H-^{13}C$ HMBC Spectrum
(DMSO-d_6, 400 MHz)

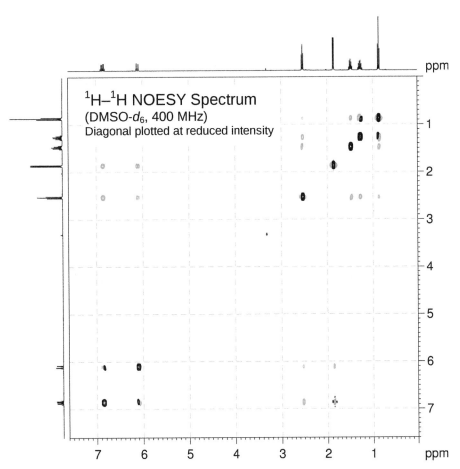

$^1H-^1H$ NOESY Spectrum
(DMSO-d_6, 400 MHz)
Diagonal plotted at reduced intensity

Problem 12

The ^{1}H and ^{13}C{^{1}H} NMR spectra of 3-nitrobenzaldehyde ($C_7H_5NO_3$) recorded in CDCl$_3$ solution at 298 K and 500 MHz are given below.

The ^{1}H NMR spectrum has signals at δ 7.82 (H$_5$), 8.28 (H$_6$), 8.51 (H$_4$), 8.73 (H$_2$) and 10.15 (H$_7$) ppm.

The ^{13}C{^{1}H} NMR spectrum has signals at δ 124.4 (C$_2$), 128.6 (C$_4$), 130.5 (C$_5$), 134.8 (C$_6$), 137.5 (C$_1$), 148.8 (C$_3$) and 189.9 (C$_7$) ppm.

Also given on the following pages are the ^{1}H–^{1}H COSY, ^{1}H–^{13}C me-HSQC, ^{1}H–^{13}C HMBC, ^{1}H–^{1}H NOESY and INADEQUATE spectra. For each 2D spectrum, indicate which correlation gives rise to each cross-peak by placing an appropriate label in the box provided.

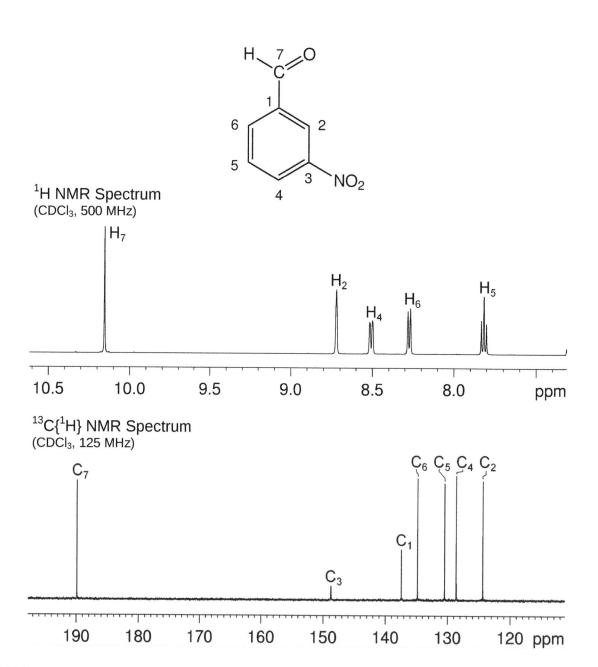

^{1}H NMR Spectrum
(CDCl$_3$, 500 MHz)

^{13}C{^{1}H} NMR Spectrum
(CDCl$_3$, 125 MHz)

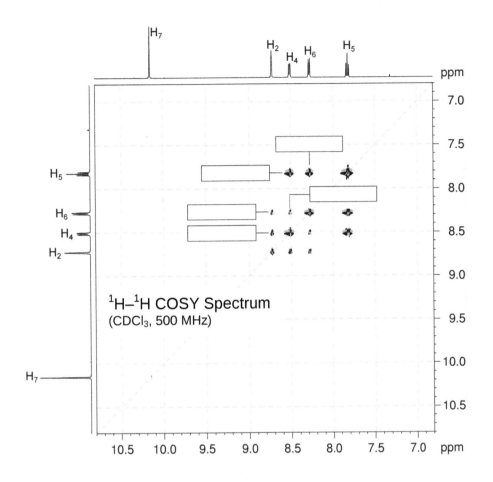

$^1H–^1H$ COSY Spectrum
(CDCl$_3$, 500 MHz)

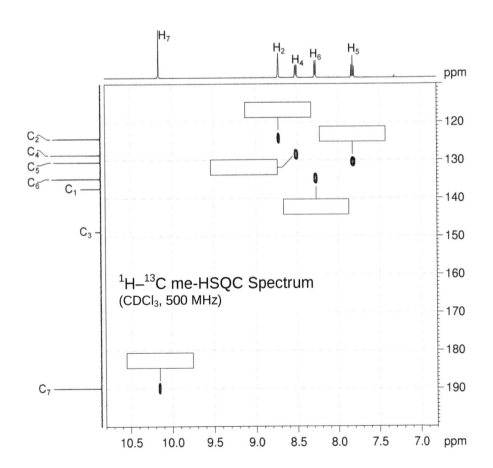

$^1H–^{13}C$ me-HSQC Spectrum
(CDCl$_3$, 500 MHz)

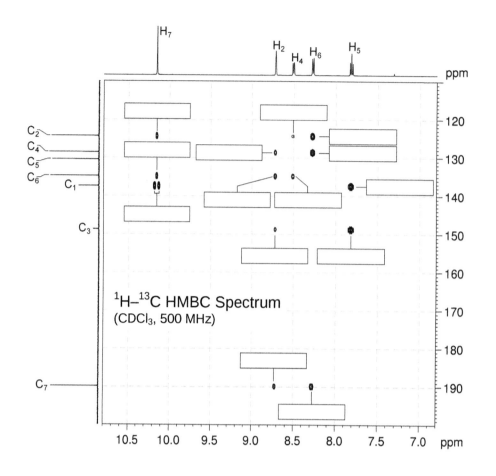

$^1H-^{13}C$ HMBC Spectrum
(CDCl$_3$, 500 MHz)

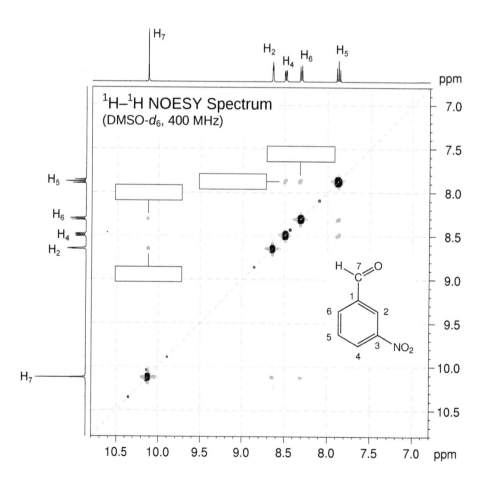

$^1H-^1H$ NOESY Spectrum
(DMSO-d_6, 400 MHz)

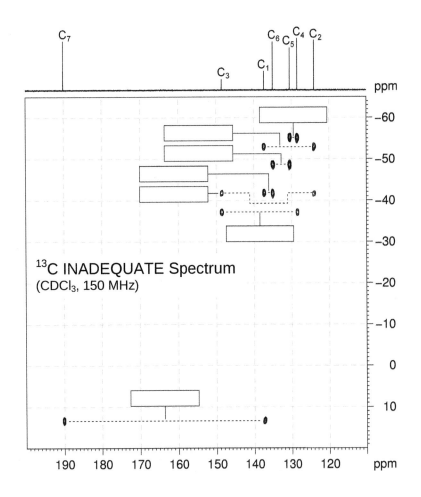

^{13}C INADEQUATE Spectrum
(CDCl$_3$, 150 MHz)

Problem 13

The ^{1}H and ^{13}C$\{^1$H$\}$ NMR spectra of 3-iodotoluene (C_7H_6I) recorded in CDCl$_3$ solution at 298 K and 600 MHz are given below.

The ^{1}H NMR spectrum has signals at δ 2.28 (H$_7$), 6.96 (H$_5$), 7.11 (H$_6$), 7.48 (H$_4$) and 7.53 (H$_2$) ppm.

The ^{13}C$\{^1$H$\}$ NMR spectrum has signals at δ 21.0, 94.3, 128.3, 129.9, 134.4, 138.0 and 140.2 ppm.

Use the me-HSQC spectrum to assign the protonated carbon signals, and then use this information to produce a schematic HMBC spectrum, showing where all of the cross-peaks would be.

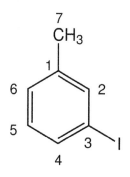

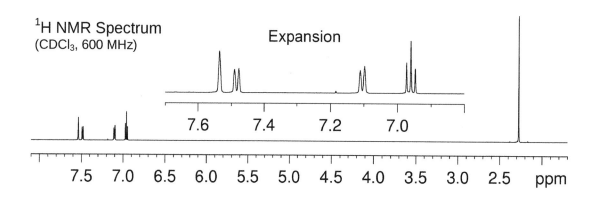

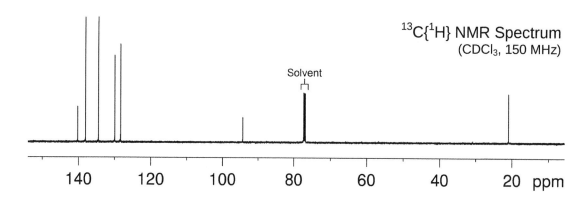

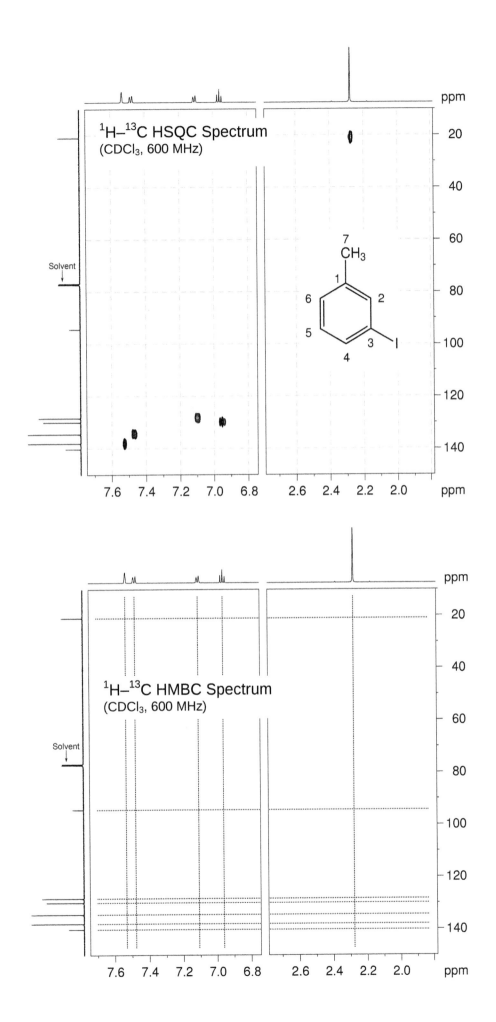

^{1}H–^{13}C HSQC Spectrum
(CDCl$_3$, 600 MHz)

Solvent

^{1}H–^{13}C HMBC Spectrum
(CDCl$_3$, 600 MHz)

Solvent

105

Problem 14

The ^{1}H and ^{13}C{^{1}H} NMR spectra of 8-hydroxy-5-nitroquinoline ($C_9H_6N_2O_3$) recorded in DMSO-d_6 solution at 298 K and 400 MHz are given below.

The ^{1}H NMR spectrum has signals at δ 7.14 (H_7), 7.82 (H_3), 8.48 (H_6), 8.97 (H_2) and 9.08 (H_4) ppm. The hydroxyl proton is not shown.

The ^{13}C{^{1}H} NMR spectrum has signals at δ 110.0 (C_7), 122.5 (C_{10}), 125.2 (C_3), 129.1 (C_6), 132.4 (C_4), 135.0 (C_5), 137.2 (C_9), 149.1 (C_2) and 160.7 (C_8) ppm.

Also given on the following pages are the ^{1}H–^{1}H COSY, ^{1}H–^{13}C me-HSQC, ^{1}H–^{13}C HMBC and INADEQUATE spectra. For each 2D spectrum, indicate which correlation gives rise to each cross-peak by placing an appropriate label in the box provided.

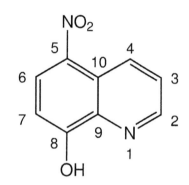

^{1}H NMR Spectrum
(DMSO-d_6, 400 MHz)

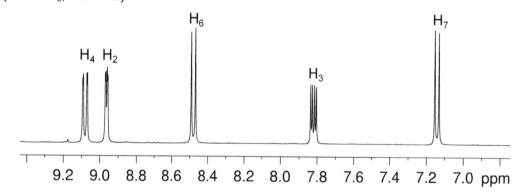

^{13}C{^{1}H} NMR Spectrum
(DMSO-d_6, 100 MHz)

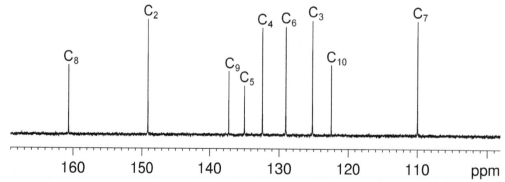

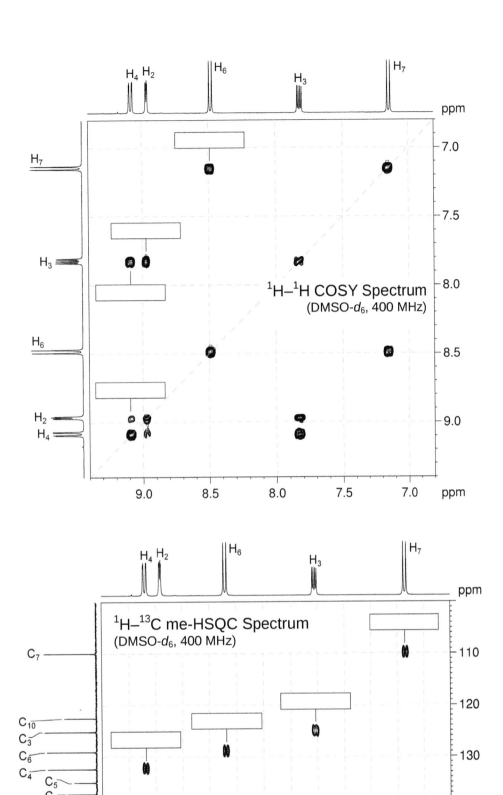

^1H–^1H COSY Spectrum
(DMSO-d_6, 400 MHz)

^1H–^{13}C me-HSQC Spectrum
(DMSO-d_6, 400 MHz)

107

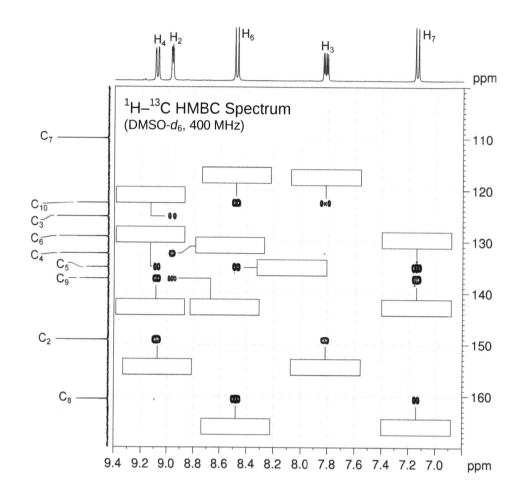

^{1}H–^{13}C HMBC Spectrum
(DMSO-d_6, 400 MHz)

H$_4$ H$_2$ H$_6$ H$_3$ H$_7$

C$_7$

C$_{10}$
C$_3$
C$_6$
C$_4$
C$_5$
C$_9$

C$_2$

C$_8$

ppm

110
120
130
140
150
160

9.4 9.2 9.0 8.8 8.6 8.4 8.2 8.0 7.8 7.6 7.4 7.2 7.0 ppm

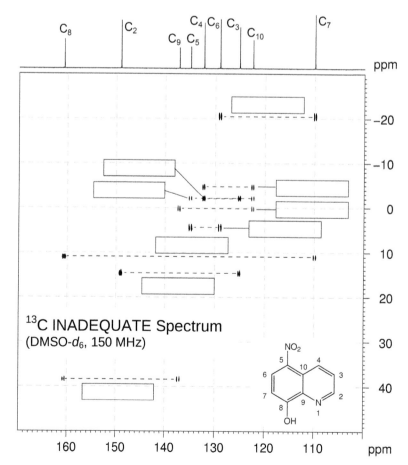

C$_8$ C$_2$ C$_4$ C$_6$ C$_3$ C$_7$
              C$_9$ C$_5$ C$_{10}$

ppm

−20
−10
0
10
20

30

40

^{13}C INADEQUATE Spectrum
(DMSO-d_6, 150 MHz)

160   150   140   130   120   110   ppm

Problem 15

Identify the following compound.
Molecular Formula: C_6H_6BrN

^1H NMR Spectrum
(CDCl$_3$, 600 MHz)

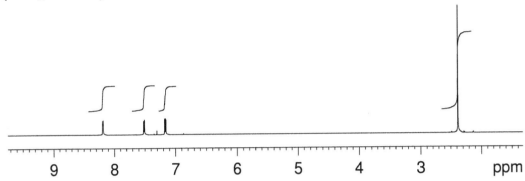

^1H NMR Expansion
(CDCl$_3$, 600 MHz)

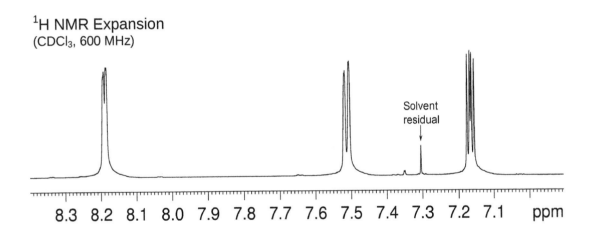

Solvent residual

$^{13}C\{^1H\}$ NMR Spectrum
(CDCl$_3$, 150 MHz)

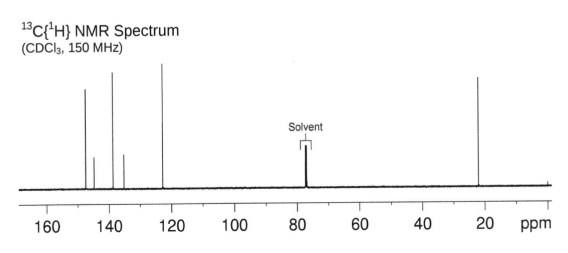

Solvent

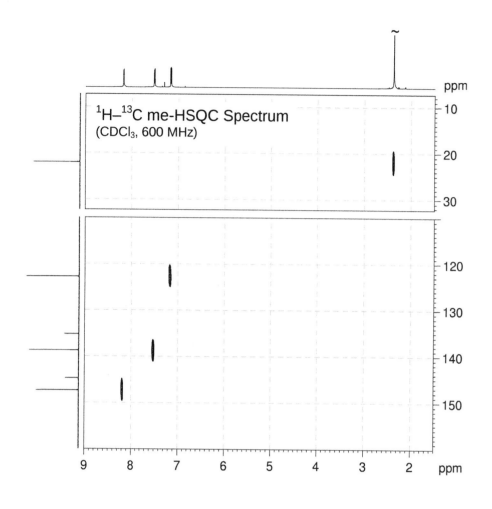

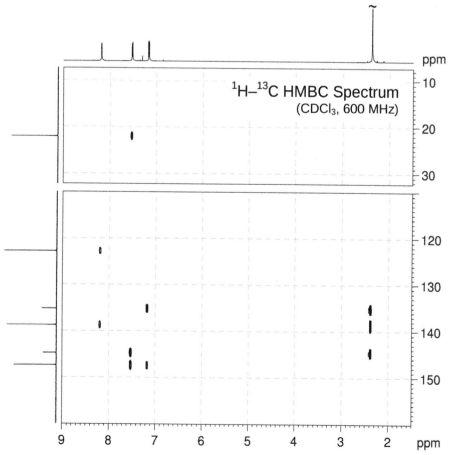

Problem 16

The ^{1}H NMR spectrum of *trans*-anethole ($C_{10}H_{12}O$) recorded in $CDCl_3$ solution at 298 K and 400 MHz is given below.

The ^{1}H NMR spectrum has signals at δ 1.82 (dd, J = 6.6, 1.7 Hz, 3H, H_1), 3.71 (s, 3H, H_8), 6.04 (dq, J = 15.8, 6.6 Hz, 1H, H_2), 6.30 (dq, J = 15.8, 1.7 Hz, 1H, H_3), 6.78 (m, 2H, H_6) and 7.21 (m, 2H, H_5) ppm.

Use this information to produce schematic diagrams of the COSY and NOESY spectra, showing where all of the cross peaks and diagonal peaks would be.

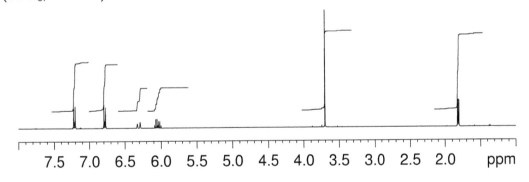

^{1}H NMR Spectrum
(CDCl$_3$, 400 MHz)

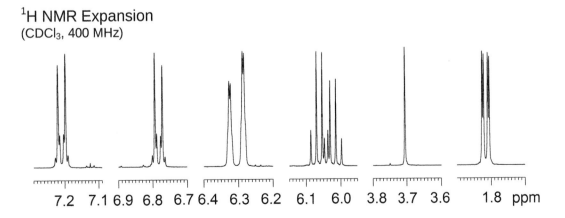

^{1}H NMR Expansion
(CDCl$_3$, 400 MHz)

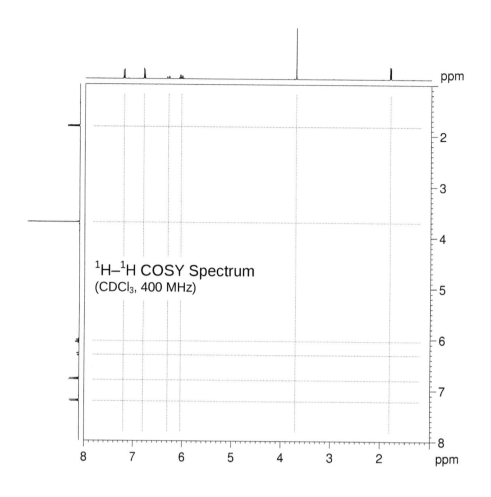

^{1}H–^{1}H COSY Spectrum
(CDCl$_3$, 400 MHz)

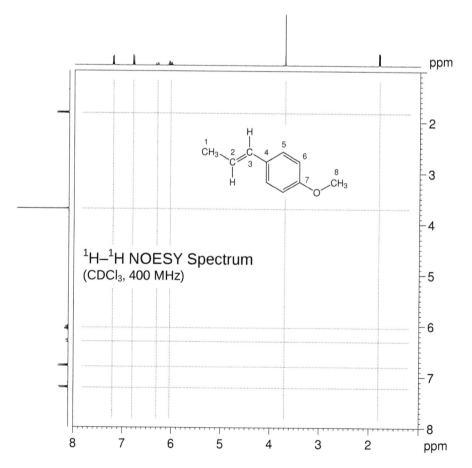

^{1}H–^{1}H NOESY Spectrum
(CDCl$_3$, 400 MHz)

Problem 17

Identify the following compound.
Molecular Formula: C_5H_{10}

^1H NMR Spectrum
(CDCl$_3$, 400 MHz)

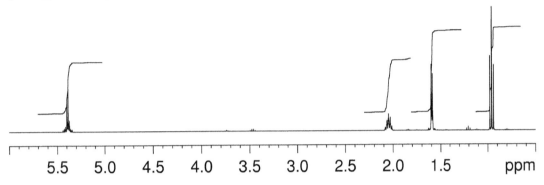

^1H NMR Expansion
(CDCl$_3$, 400 MHz)

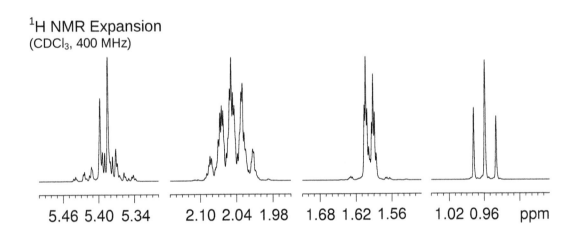

5.46 5.40 5.34   2.10 2.04 1.98   1.68 1.62 1.56   1.02 0.96   ppm

$^{13}C\{^1H\}$ NMR Spectrum
(CDCl$_3$, 100 MHz)

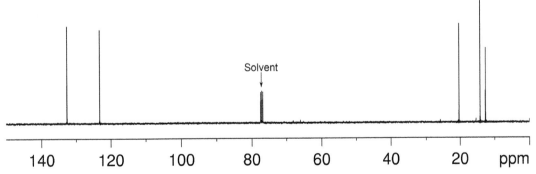

Solvent

140   120   100   80   60   40   20   ppm

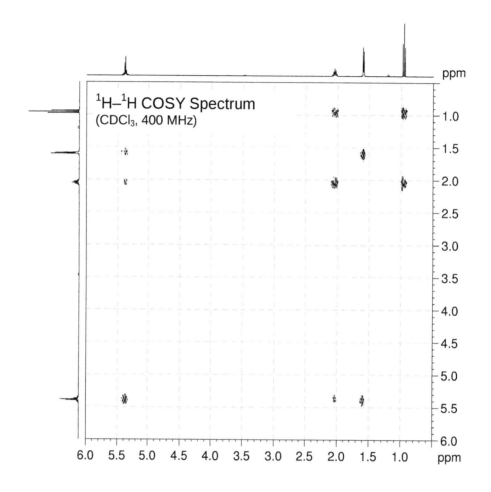

^1H–^1H COSY Spectrum
(CDCl$_3$, 400 MHz)

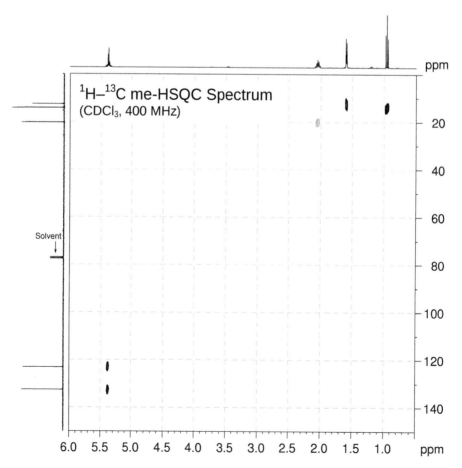

^1H–^{13}C me-HSQC Spectrum
(CDCl$_3$, 400 MHz)

Solvent

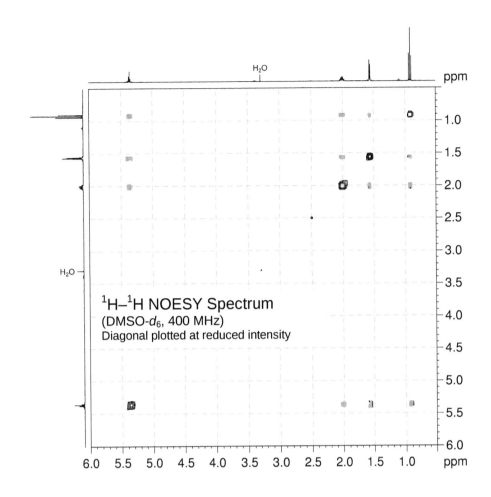

^1H–^1H NOESY Spectrum
(DMSO-d_6, 400 MHz)
Diagonal plotted at reduced intensity

115

Problem 18

Identify the following compound.
Molecular Formula: $C_{14}H_{12}O_2$
IR: 1720 cm^{-1}

^{1}H NMR Spectrum
(CDCl$_3$, 500 MHz)

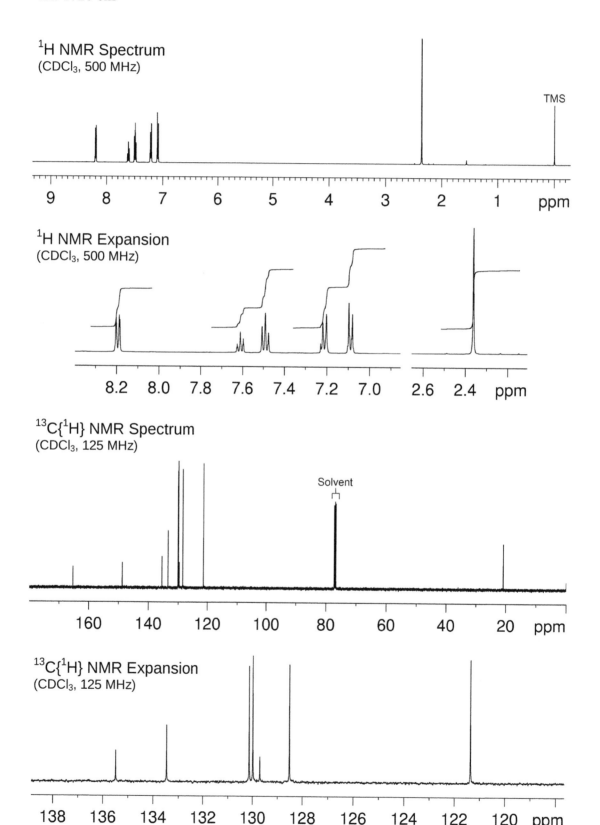

^{1}H NMR Expansion
(CDCl$_3$, 500 MHz)

^{13}C{^{1}H} NMR Spectrum
(CDCl$_3$, 125 MHz)

Solvent

^{13}C{^{1}H} NMR Expansion
(CDCl$_3$, 125 MHz)

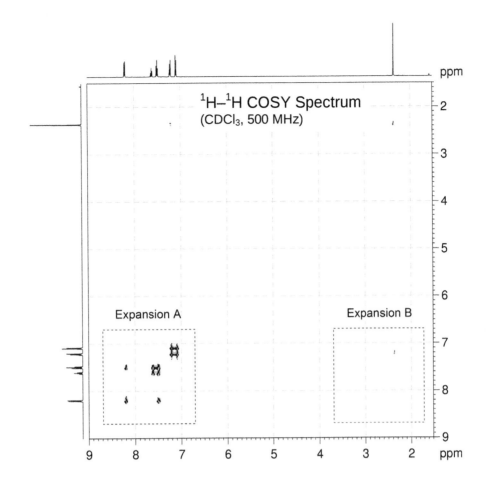

^{1}H–^{1}H COSY Spectrum
(CDCl$_3$, 500 MHz)

Expansion A

Expansion B

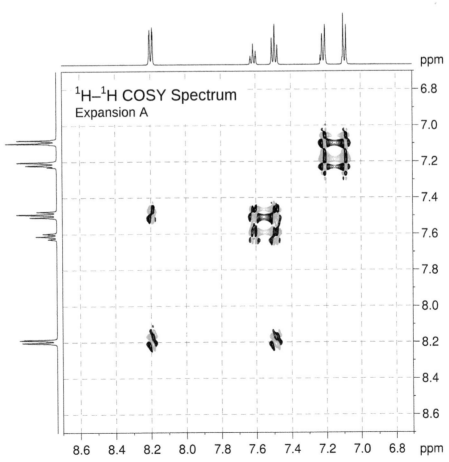

^{1}H–^{1}H COSY Spectrum
Expansion A

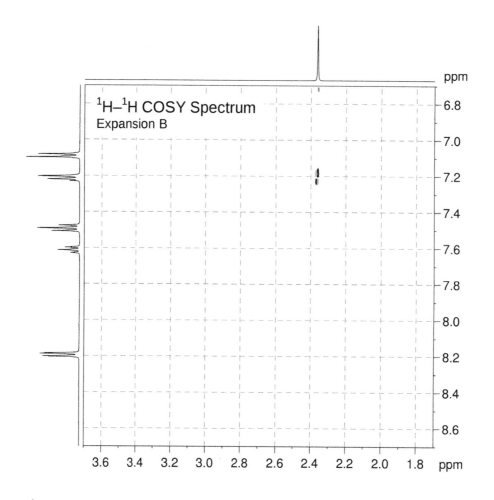

¹H–¹H COSY Spectrum
Expansion B

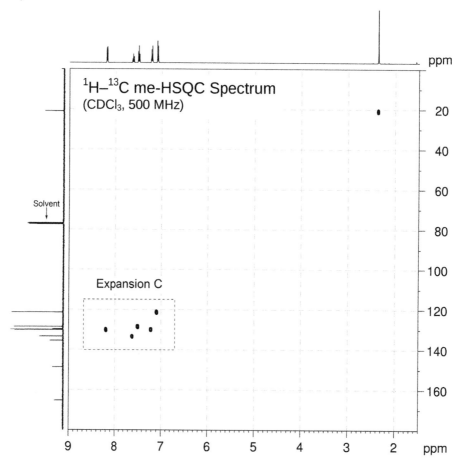

¹H–¹³C me-HSQC Spectrum
(CDCl₃, 500 MHz)

Solvent

Expansion C

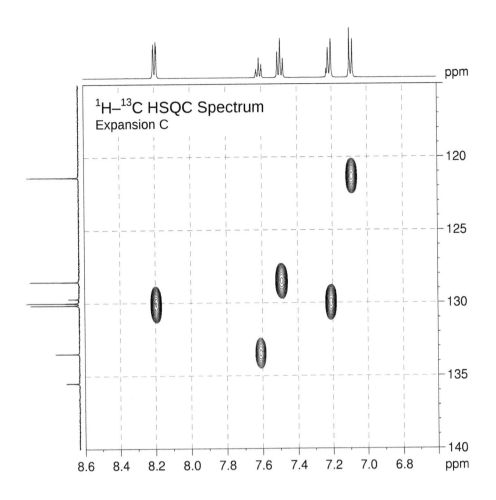

$^1H–^{13}C$ HSQC Spectrum
Expansion C

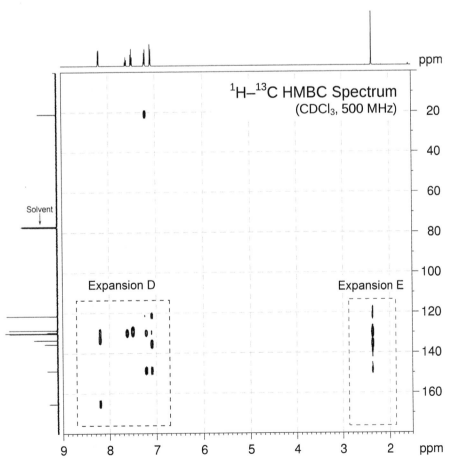

$^1H–^{13}C$ HMBC Spectrum
(CDCl$_3$, 500 MHz)

Solvent

Expansion D

Expansion E

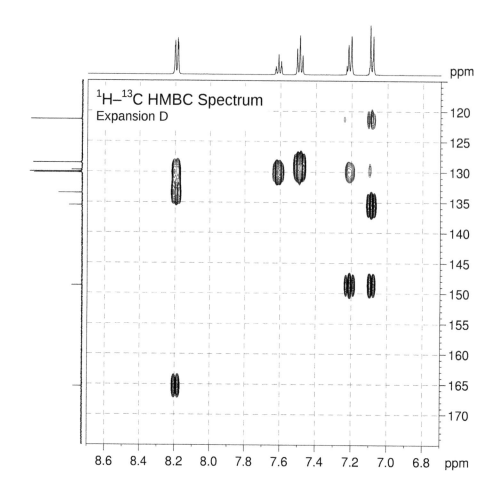

¹H–¹³C HMBC Spectrum
Expansion D

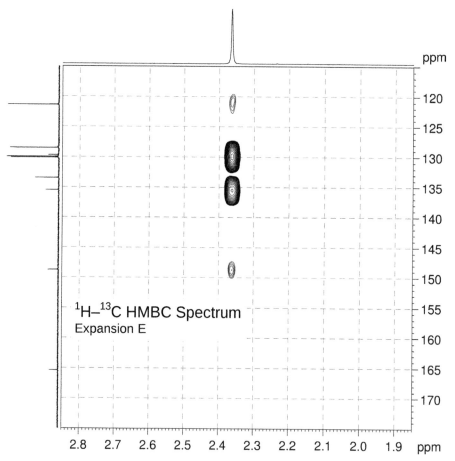

¹H–¹³C HMBC Spectrum
Expansion E

Problem 19

Identify the following compound.

Molecular Formula: $C_{14}H_{12}O_2$

IR: 1720 cm^{-1}

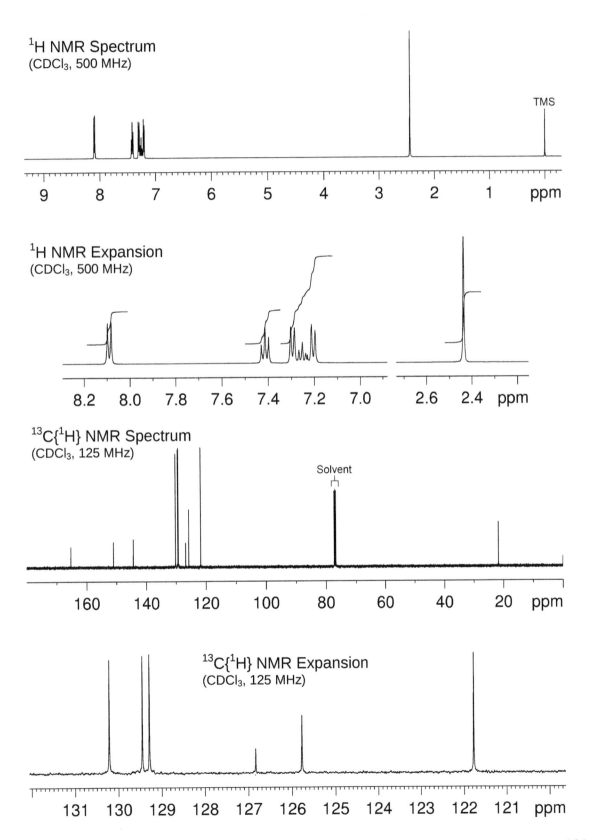

^{1}H NMR Spectrum
(CDCl$_3$, 500 MHz)

TMS

^{1}H NMR Expansion
(CDCl$_3$, 500 MHz)

^{13}C{^{1}H} NMR Spectrum
(CDCl$_3$, 125 MHz)

Solvent

^{13}C{^{1}H} NMR Expansion
(CDCl$_3$, 125 MHz)

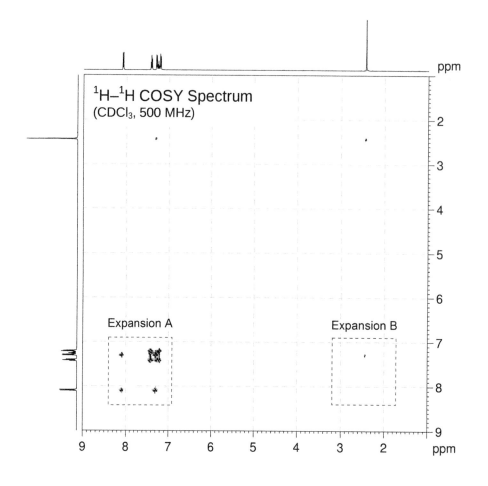

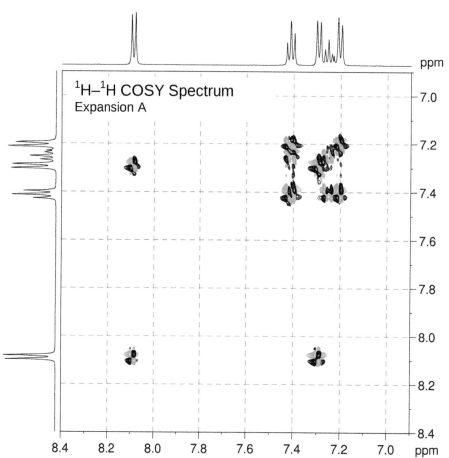

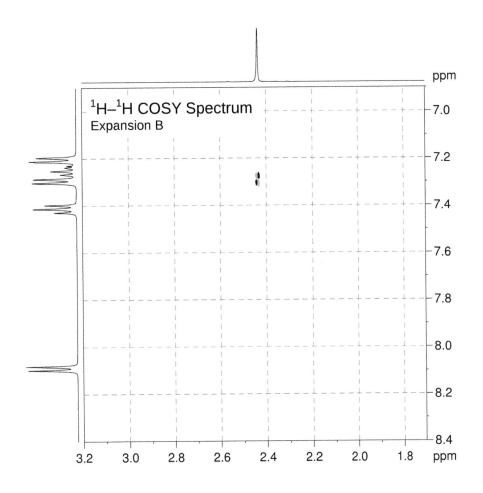

^{1}H–^{1}H COSY Spectrum
Expansion B

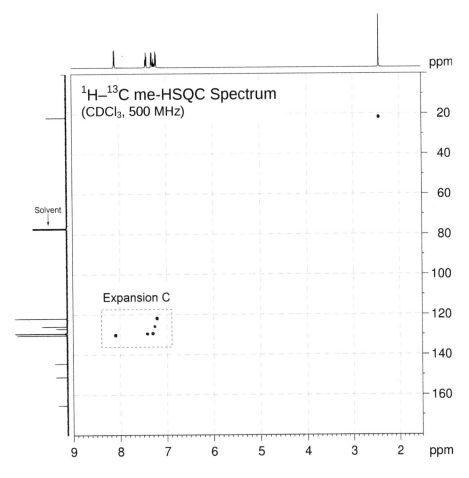

^{1}H–^{13}C me-HSQC Spectrum
(CDCl$_3$, 500 MHz)

Solvent

Expansion C

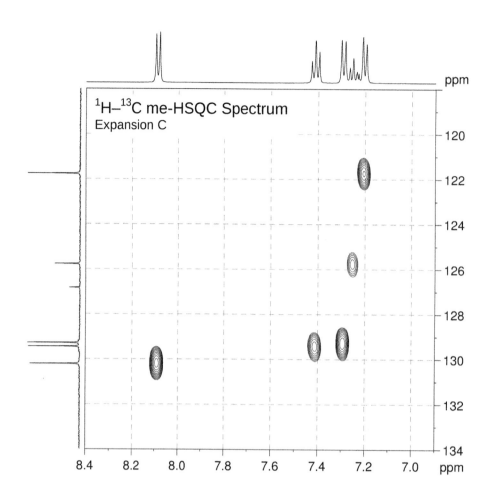

^1H–^{13}C me-HSQC Spectrum
Expansion C

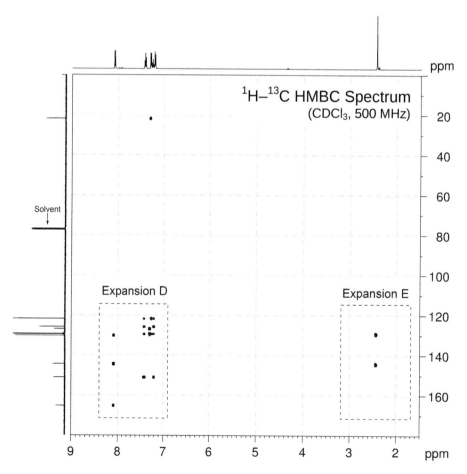

^1H–^{13}C HMBC Spectrum
(CDCl$_3$, 500 MHz)

Solvent

Expansion D

Expansion E

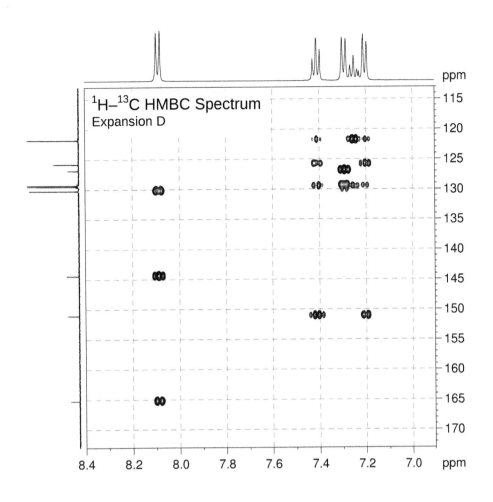

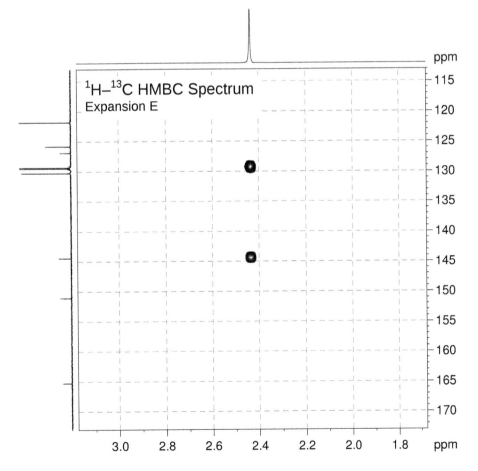

Problem 20

Identify the following compound.
Molecular Formula: $C_{14}H_{12}O_2$
IR: 1751 cm^{-1}

^{1}H NMR Spectrum
(CDCl$_3$, 500 MHz)

TMS

^{1}H NMR Expansion
(CDCl$_3$, 500 MHz)

Solvent
residual

^{13}C{^{1}H} NMR Spectrum
(CDCl$_3$, 100 MHz)

Solvent

^{13}C{^{1}H} NMR Expansion
(CDCl$_3$, 100 MHz)

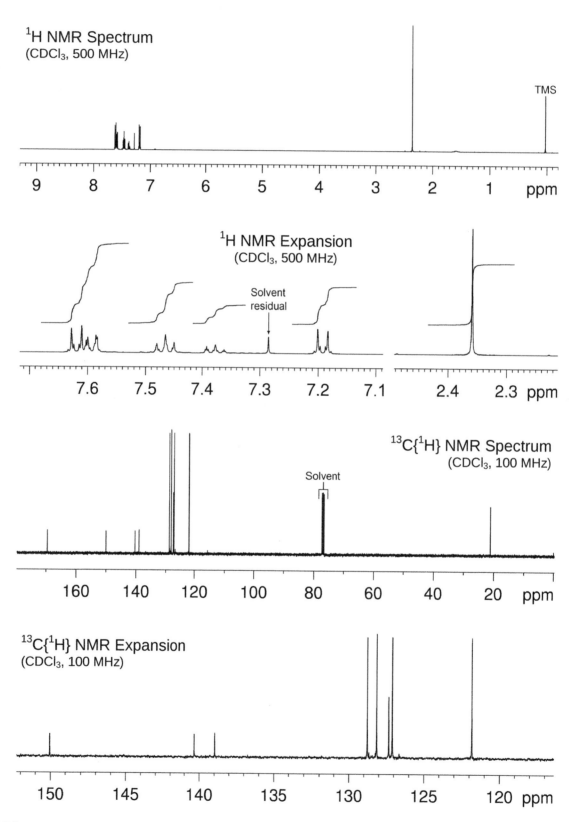

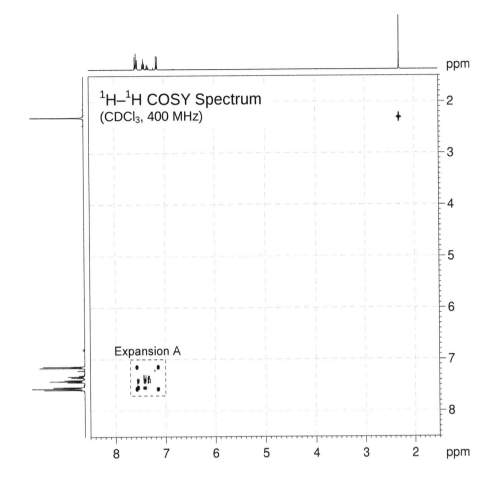

^1H–^1H COSY Spectrum
(CDCl₃, 400 MHz)

Expansion A

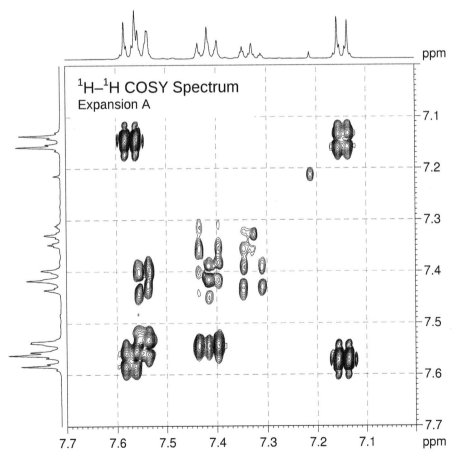

^1H–^1H COSY Spectrum
Expansion A

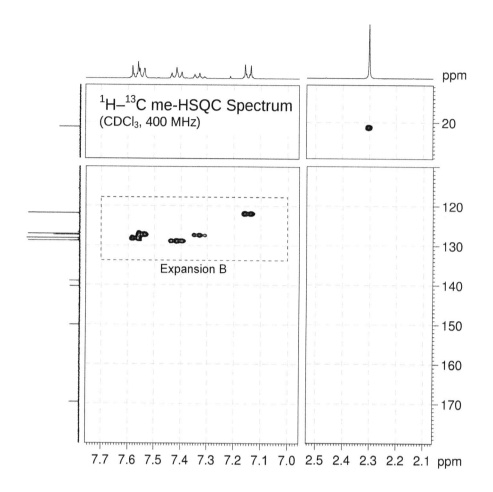

$^1H-^{13}C$ me-HSQC Spectrum
(CDCl$_3$, 400 MHz)

Expansion B

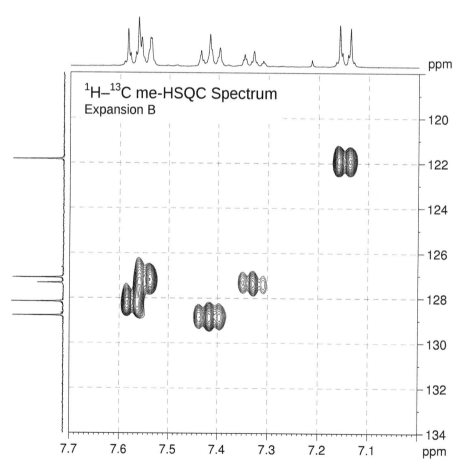

$^1H-^{13}C$ me-HSQC Spectrum
Expansion B

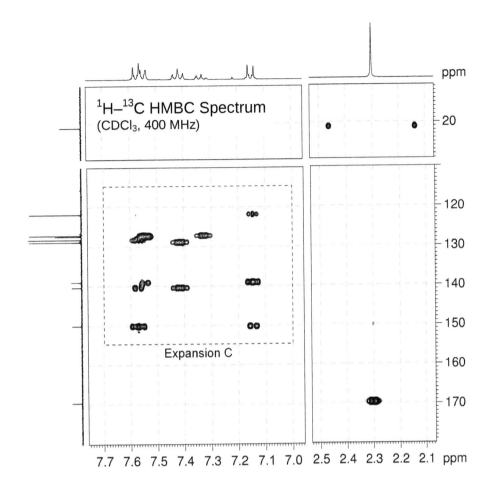

^1H–^{13}C HMBC Spectrum
(CDCl$_3$, 400 MHz)

Expansion C

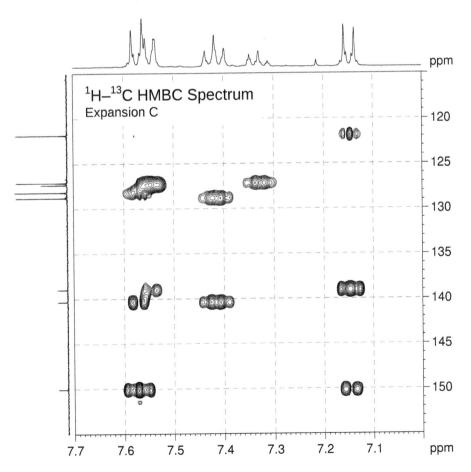

^1H–^{13}C HMBC Spectrum
Expansion C

Problem 21

Identify the following compound.
Molecular Formula: $C_{14}H_{12}O_2$
IR: 1678 cm^{-1}

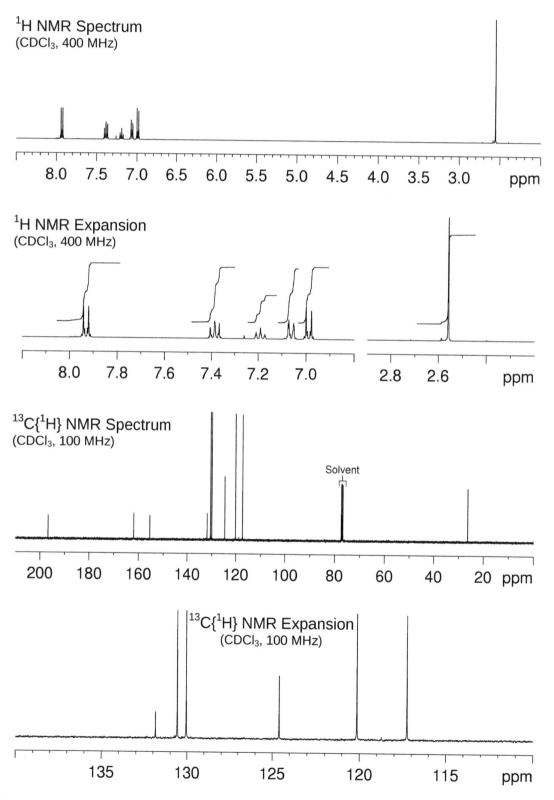

^{1}H NMR Spectrum
(CDCl$_3$, 400 MHz)

^{1}H NMR Expansion
(CDCl$_3$, 400 MHz)

^{13}C{^{1}H} NMR Spectrum
(CDCl$_3$, 100 MHz)

Solvent

^{13}C{^{1}H} NMR Expansion
(CDCl$_3$, 100 MHz)

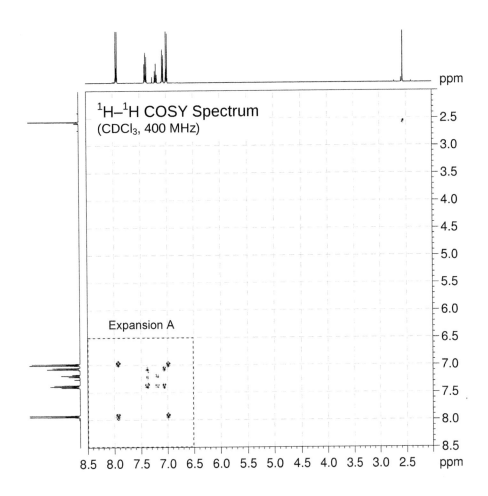

^{1}H–^{1}H COSY Spectrum
(CDCl$_3$, 400 MHz)

Expansion A

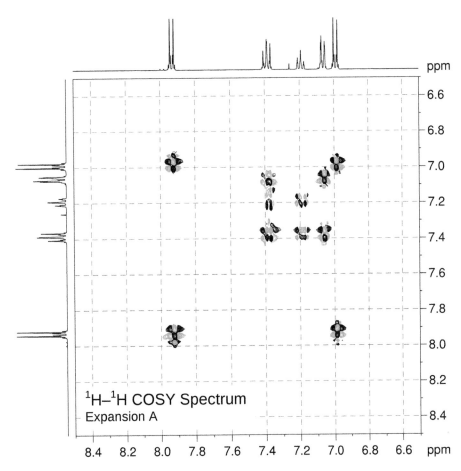

^{1}H–^{1}H COSY Spectrum
Expansion A

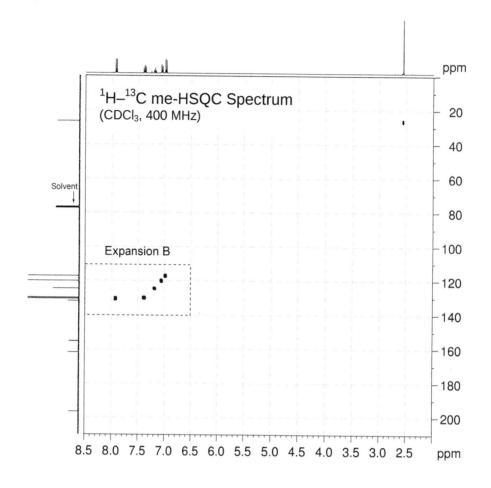

$^1H-^{13}C$ me-HSQC Spectrum
(CDCl$_3$, 400 MHz)

Expansion B

Solvent

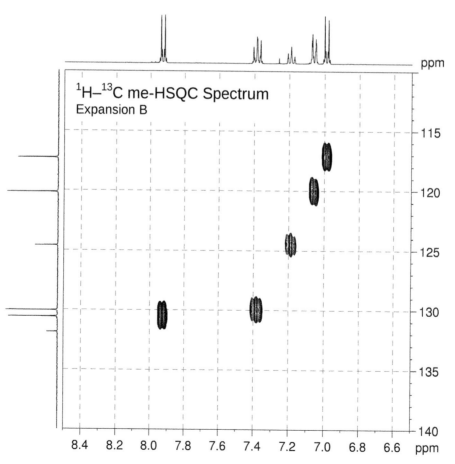

$^1H-^{13}C$ me-HSQC Spectrum
Expansion B

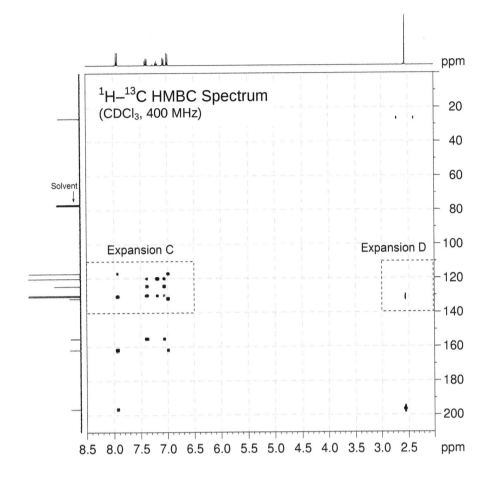

^{1}H–^{13}C HMBC Spectrum
(CDCl$_3$, 400 MHz)

Expansion C

Expansion D

Solvent

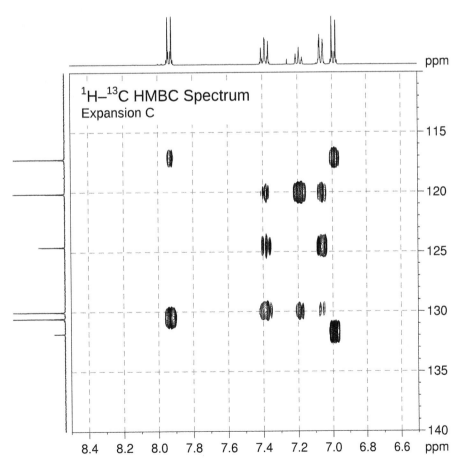

^{1}H–^{13}C HMBC Spectrum
Expansion C

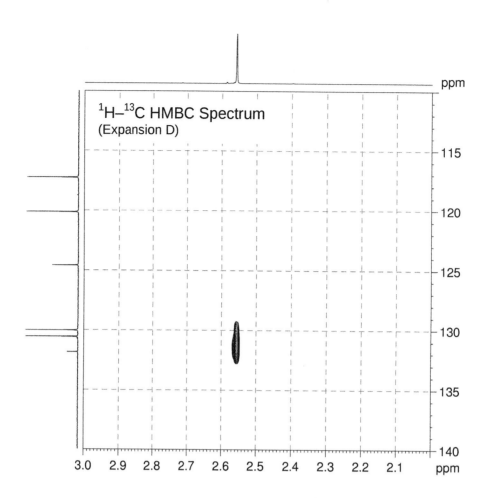

^{1}H–^{13}C HMBC Spectrum
(Expansion D)

Problem 22

Identify the following compound.

Molecular Formula: $C_{12}H_{16}O$

IR Spectrum: 1686 cm^{-1}

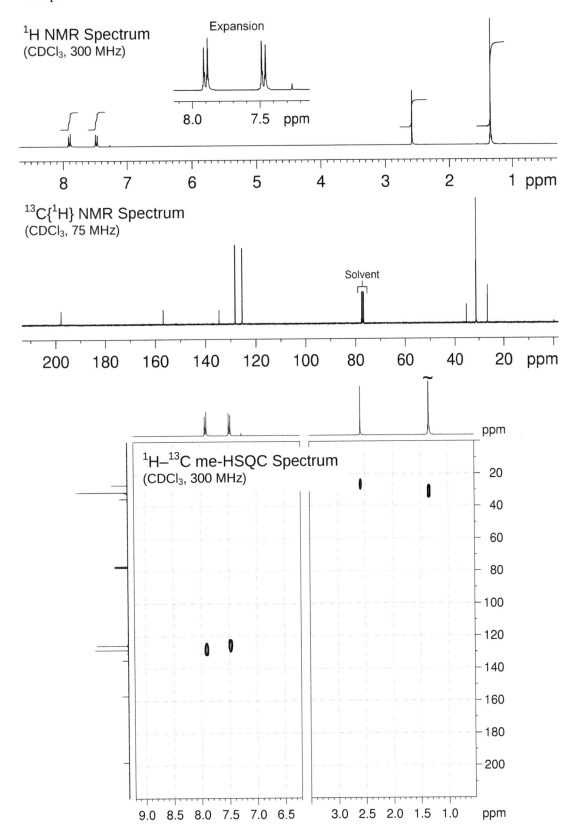

¹H NMR Spectrum
(CDCl₃, 300 MHz)

Expansion

8.0   7.5   ppm

8   7   6   5   4   3   2   1 ppm

¹³C{¹H} NMR Spectrum
(CDCl₃, 75 MHz)

Solvent

200 180 160 140 120 100 80 60 40 20 ppm

¹H–¹³C me-HSQC Spectrum
(CDCl₃, 300 MHz)

ppm

20

40

60

80

100

120

140

160

180

200

9.0 8.5 8.0 7.5 7.0 6.5   3.0 2.5 2.0 1.5 1.0 ppm

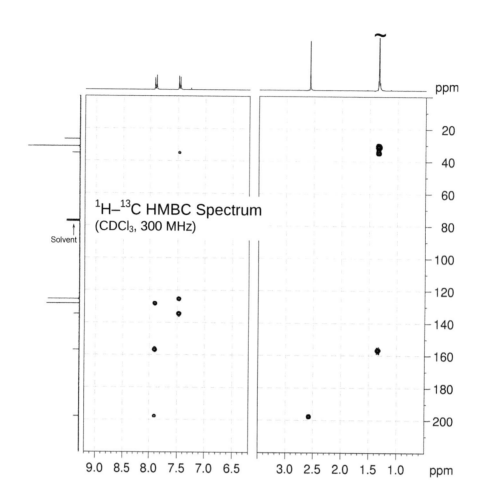

$^1H-^{13}C$ HMBC Spectrum
(CDCl$_3$, 300 MHz)

Solvent

136

Problem 23

Identify the following compound.
Molecular Formula: $C_{12}H_{16}O$
IR Spectrum: 1685 cm^{-1}

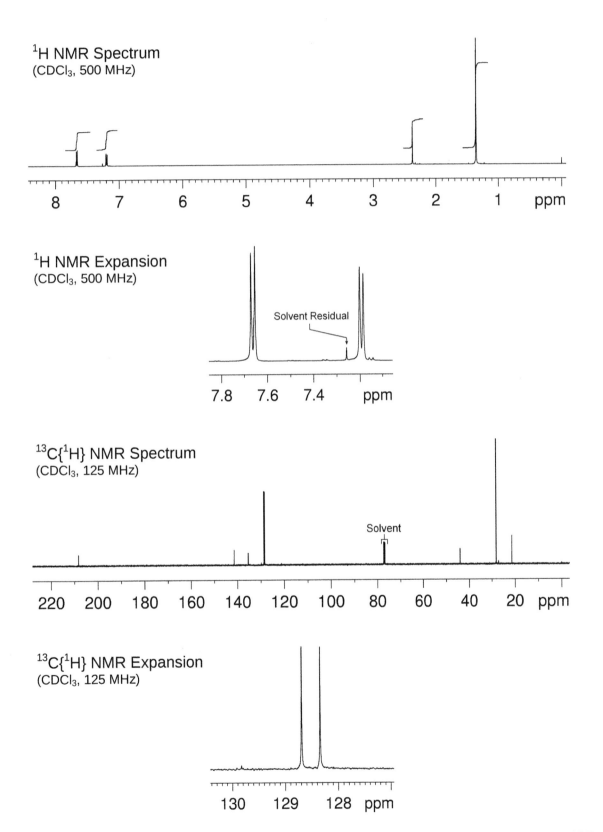

^{1}H NMR Spectrum
(CDCl$_3$, 500 MHz)

^{1}H NMR Expansion
(CDCl$_3$, 500 MHz)

Solvent Residual

^{13}C{^{1}H} NMR Spectrum
(CDCl$_3$, 125 MHz)

Solvent

^{13}C{^{1}H} NMR Expansion
(CDCl$_3$, 125 MHz)

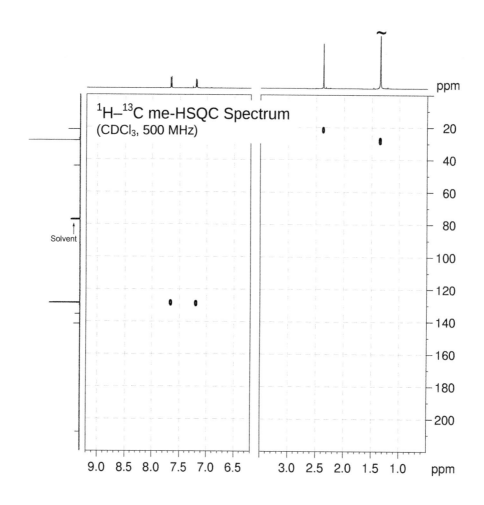

¹H–¹³C me-HSQC Spectrum
(CDCl₃, 500 MHz)

Solvent

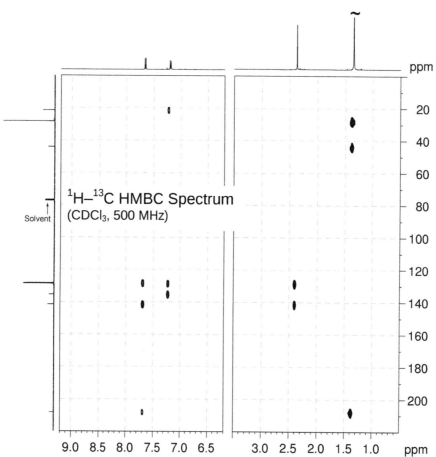

¹H–¹³C HMBC Spectrum
(CDCl₃, 500 MHz)

Solvent

Problem 24

Identify the following compound.

Molecular Formula: $C_{10}H_{12}O$

IR: 3320 (br), 1600 (w), 1492 (m), 1445 (m) cm^{-1}

^{1}H NMR Spectrum
(DMSO-d_6, 500 MHz)

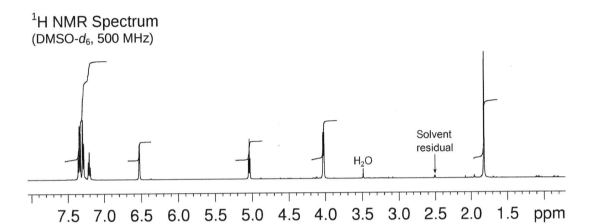

^{1}H NMR Expansion
(DMSO-d_6, 500 MHz)

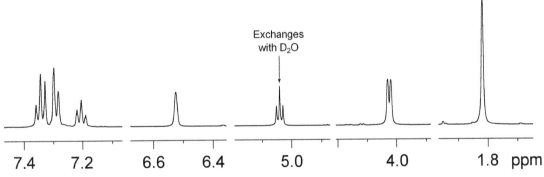

^{13}C{^{1}H} NMR Spectrum
(DMSO-d_6, 125 MHz)

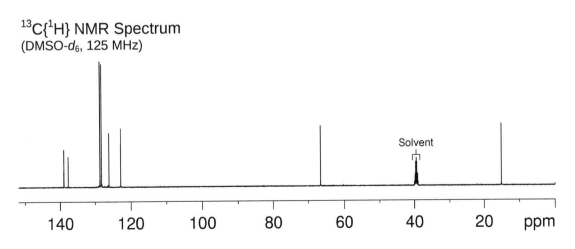

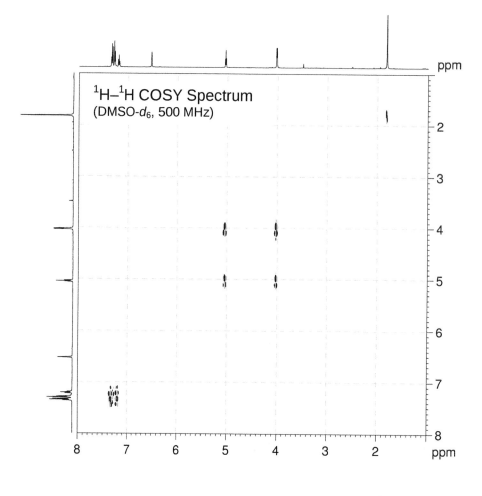

^1H–^1H COSY Spectrum
(DMSO-d_6, 500 MHz)

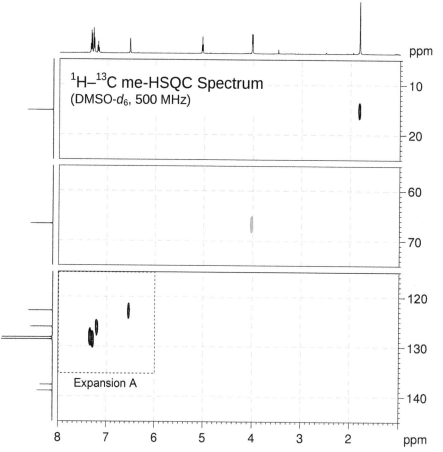

^1H–^{13}C me-HSQC Spectrum
(DMSO-d_6, 500 MHz)

Expansion A

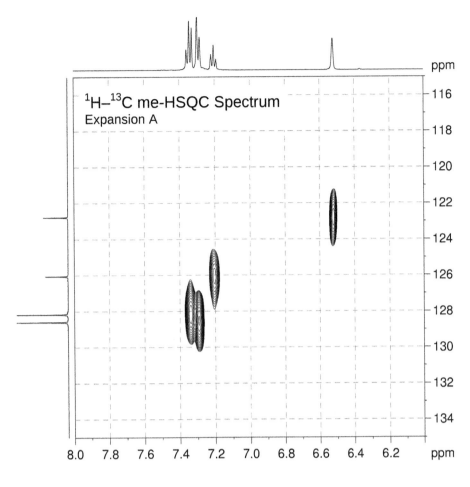

$^1H–^{13}C$ me-HSQC Spectrum
Expansion A

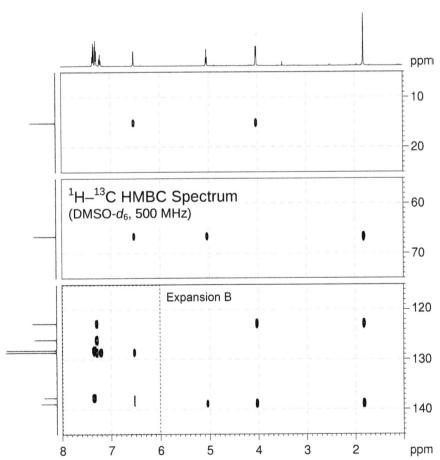

$^1H–^{13}C$ HMBC Spectrum
(DMSO-d_6, 500 MHz)

Expansion B

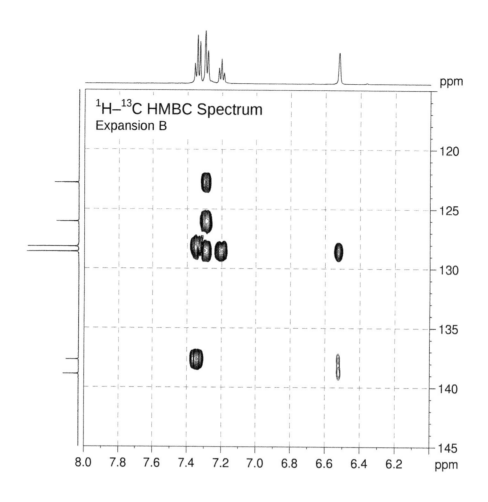

$^1H–^{13}C$ HMBC Spectrum
Expansion B

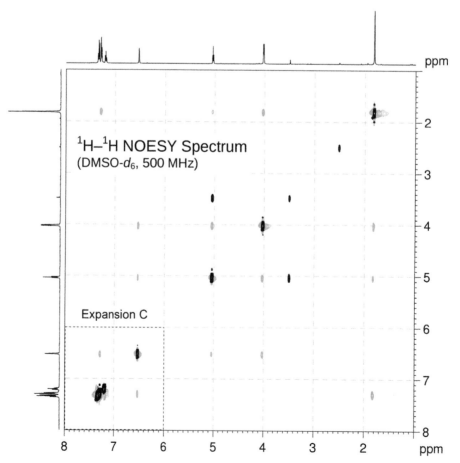

$^1H–^1H$ NOESY Spectrum
(DMSO-d_6, 500 MHz)

Expansion C

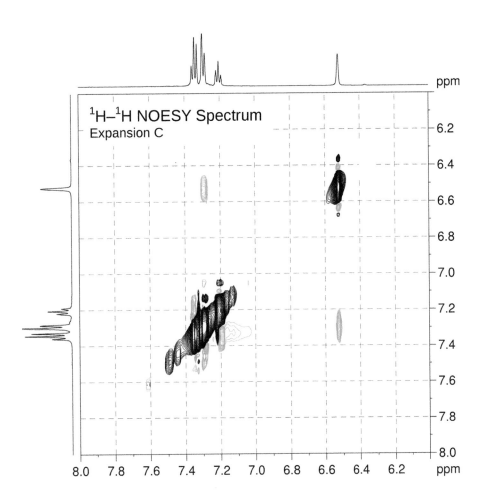

^{1}H–^{1}H NOESY Spectrum
Expansion C

143

Problem 25

Identify the following compound.
Molecular Formula: $C_{10}H_{12}O_3$
IR: 1720 cm^{-1}

^{1}H NMR Spectrum
(CDCl$_3$, 500 MHz)

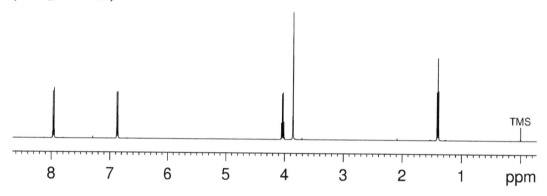

^{1}H NMR Expansion
(CDCl$_3$, 500 MHz)

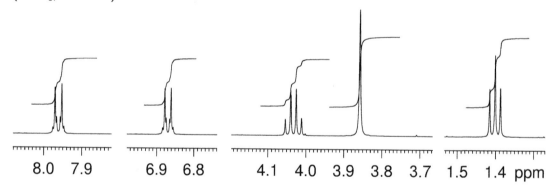

^{13}C{^{1}H} NMR Spectrum
(CDCl$_3$, 125 MHz)

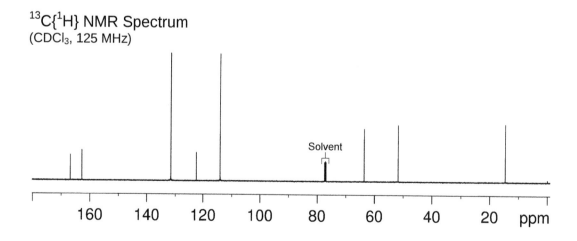

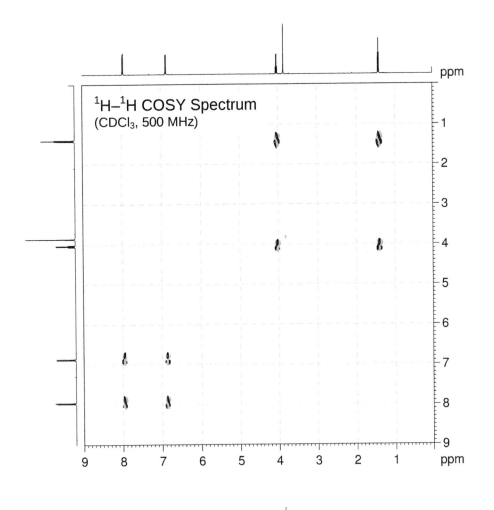

^{1}H–^{1}H COSY Spectrum
(CDCl$_3$, 500 MHz)

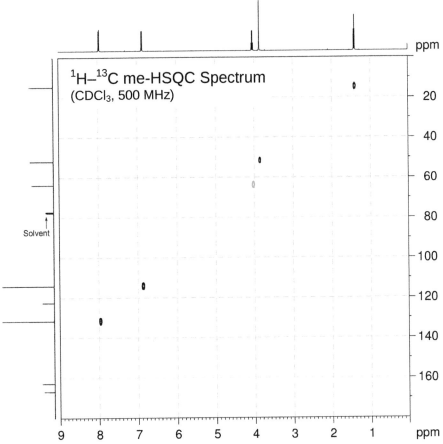

^{1}H–^{13}C me-HSQC Spectrum
(CDCl$_3$, 500 MHz)

Solvent

145

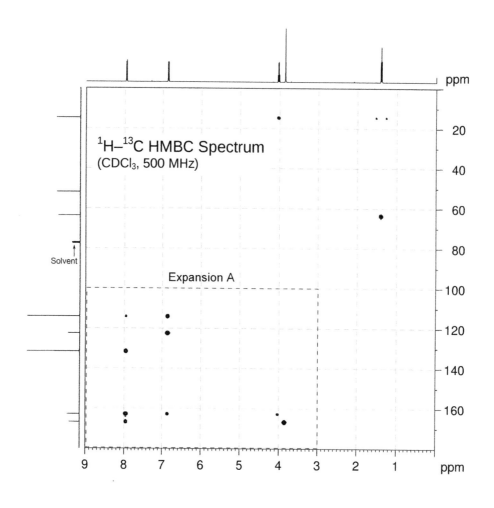

^1H–^{13}C HMBC Spectrum
(CDCl$_3$, 500 MHz)

Expansion A

Solvent

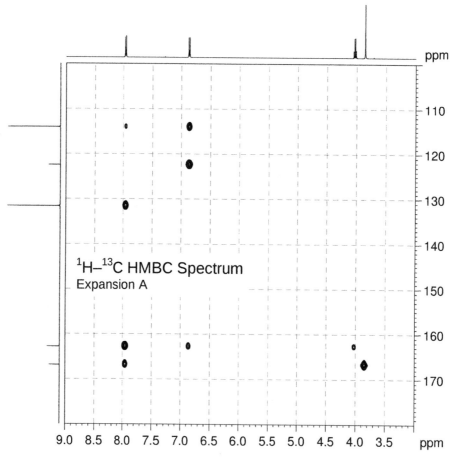

^1H–^{13}C HMBC Spectrum
Expansion A

Problem 26

Identify the following compound.
Molecular Formula: $C_{11}H_{14}O_2$
IR: 1739 cm^{-1}

^{1}H NMR Spectrum
(CDCl$_3$, 500 MHz)

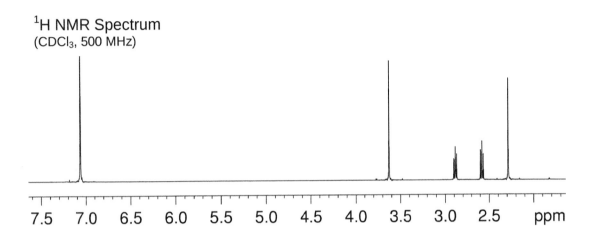

^{1}H NMR Expansion
(CDCl$_3$, 500 MHz)

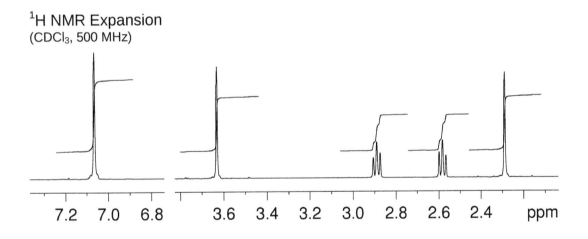

^{13}C{^{1}H} NMR Spectrum
(CDCl$_3$, 125 MHz)

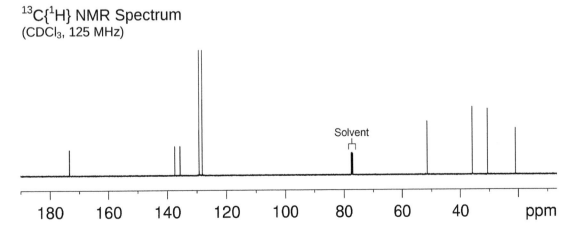

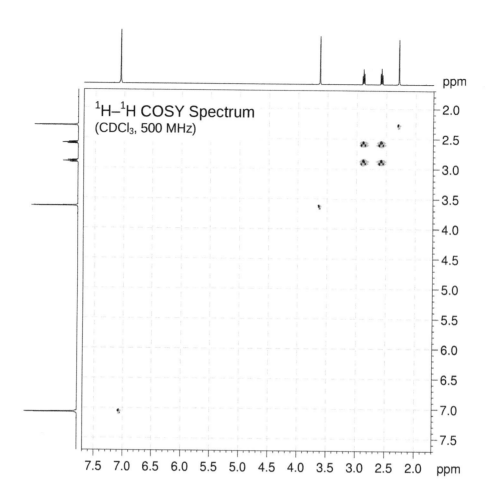

¹H–¹H COSY Spectrum
(CDCl₃, 500 MHz)

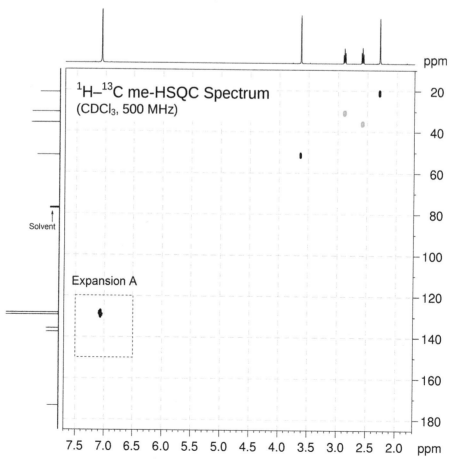

¹H–¹³C me-HSQC Spectrum
(CDCl₃, 500 MHz)

Solvent

Expansion A

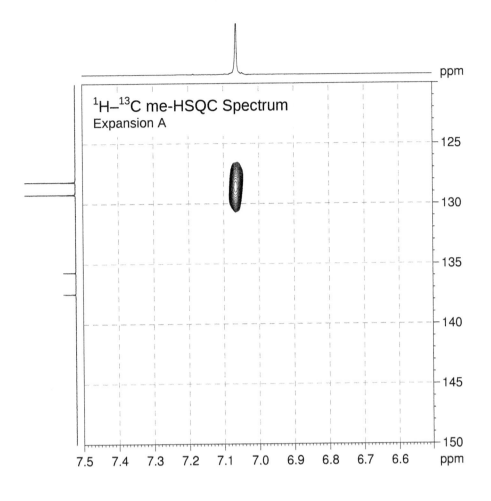

^{1}H–^{13}C me-HSQC Spectrum
Expansion A

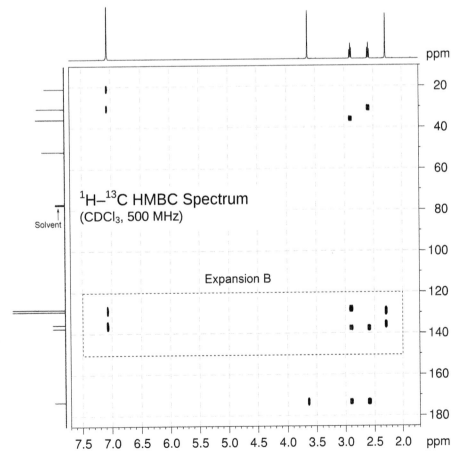

^{1}H–^{13}C HMBC Spectrum
(CDCl$_3$, 500 MHz)

Expansion B

Solvent

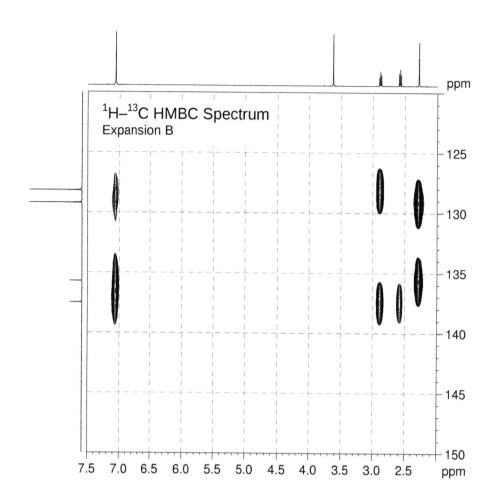

^{1}H–^{13}C HMBC Spectrum
Expansion B

150

Problem 27

Identify the following compound.

Molecular Formula: $C_{11}H_{14}O_2$

IR: 1715 cm^{-1}

Draw a labelled structure, and use the me-HSQC and INADEQUATE spectra to assign each ^{1}H and ^{13}C resonance to the corresponding nucleus in the structure.

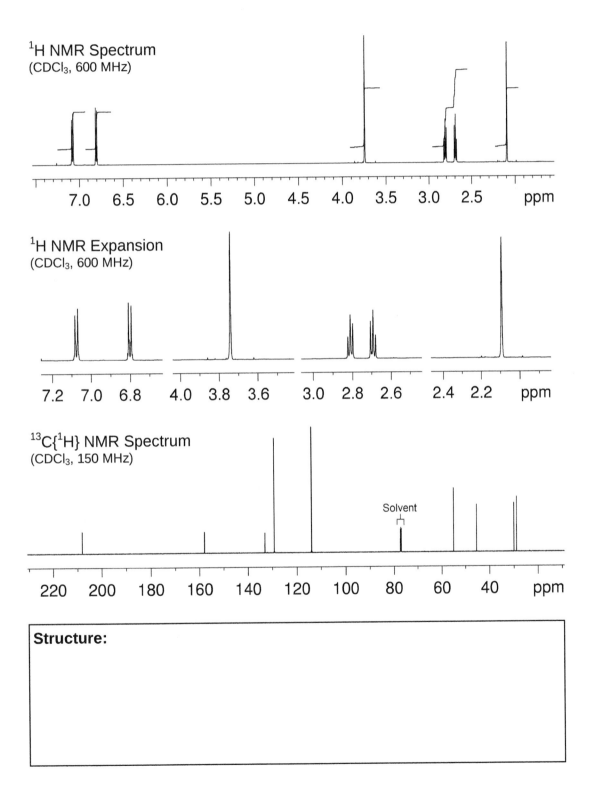

^{1}H NMR Spectrum
(CDCl$_3$, 600 MHz)

^{1}H NMR Expansion
(CDCl$_3$, 600 MHz)

^{13}C{^{1}H} NMR Spectrum
(CDCl$_3$, 150 MHz)

Solvent

Structure:

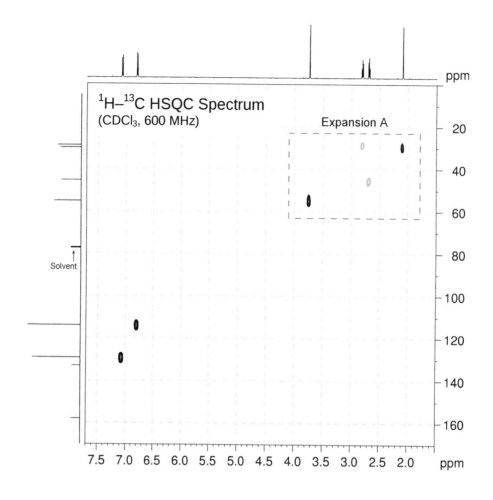

¹H–¹³C HSQC Spectrum
(CDCl₃, 600 MHz)

Expansion A

Solvent

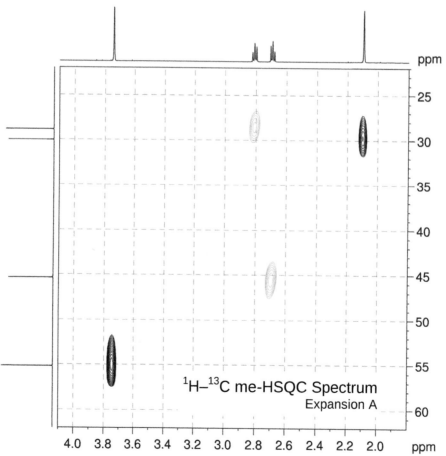

¹H–¹³C me-HSQC Spectrum
Expansion A

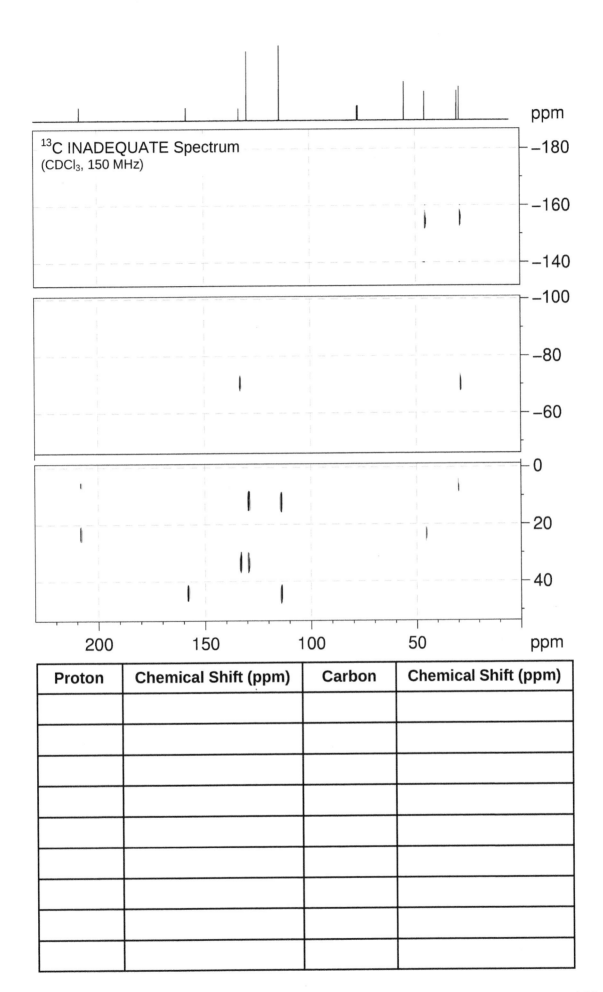

Proton	Chemical Shift (ppm)	Carbon	Chemical Shift (ppm)

153

Problem 28

The ^{1}H and ^{13}C{^{1}H} NMR spectra of ethyl 6-bromohexanoate ($C_8H_{15}BrO_2$) recorded in CDCl$_3$ solution at 298 K and 500 MHz are given below. The ^{1}H NMR spectrum has signals at δ 1.25, 1.48, 1.65, 1.87, 2.31, 3.41 and 4.12 ppm. The ^{13}C{^{1}H} NMR spectrum has signals at δ 14.3, 24.0, 27.6, 32.4, 33.5, 34.0, 60.2 and 173.3 ppm.

The 2D ^{1}H–^{1}H COSY and the multiplicity-edited ^{1}H–^{13}C HSQC spectra are given on the facing page. From the COSY spectrum, assign the proton spectrum, then use this information to assign the ^{13}C{^{1}H} spectrum.

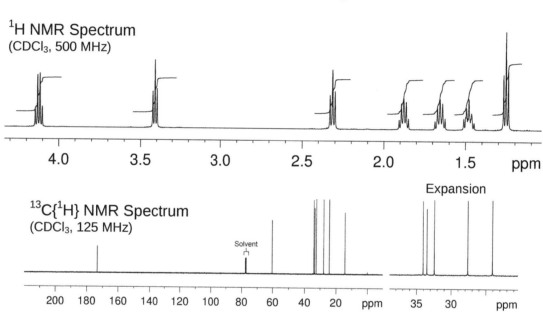

Proton	Chemical Shift (ppm)	Carbon	Chemical Shift (ppm)
H$_1$		C$_1$	
H$_2$		C$_2$	
H$_3$		C$_3$	
H$_4$		C$_4$	
H$_5$		C$_5$	
		C$_6$	
H$_7$		C$_7$	
H$_8$		C$_8$	

154

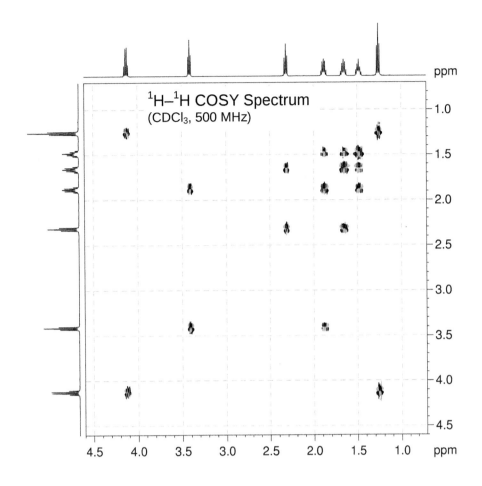

^{1}H–^{1}H COSY Spectrum
(CDCl$_3$, 500 MHz)

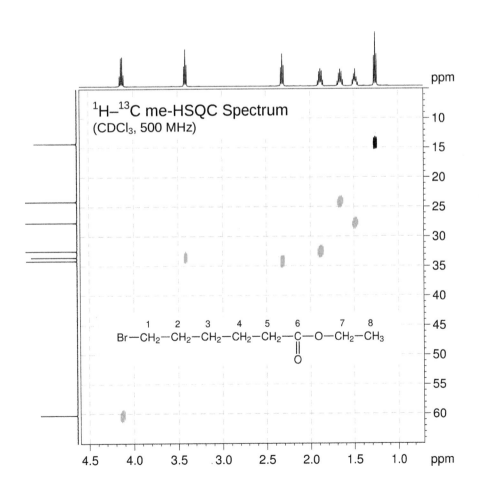

^{1}H–^{13}C me-HSQC Spectrum
(CDCl$_3$, 500 MHz)

$$\underset{1}{\text{Br}}-\underset{1}{\text{CH}_2}-\underset{2}{\text{CH}_2}-\underset{3}{\text{CH}_2}-\underset{4}{\text{CH}_2}-\underset{5}{\text{CH}_2}-\underset{6}{\overset{\displaystyle O}{\underset{\|}{\text{C}}}}-\text{O}-\underset{7}{\text{CH}_2}-\underset{8}{\text{CH}_3}$$

Problem 29

The ^1H and $^{13}C\{^1H\}$ NMR spectra of piperonal ($C_8H_6O_3$) recorded in $CDCl_3$ solution at 298 K and 600 MHz are given below.

The ^1H NMR spectrum has signals at δ 6.07 (s, 2H, H_7), 6.92 (d, $^3J_{HH}$ = 7.9 Hz, 1H, H_1), 7.31 (d, $^4J_{HH}$ = 1.5 Hz, 1H, H_4), 7.40 (dd, $^3J_{HH}$ = 7.9 Hz, $^4J_{HH}$ = 1.5 Hz, 1H, H_6) and 9.80 (s, 1H, H_8) ppm.

The $^{13}C\{^1H\}$ NMR spectrum has signals at δ 102.1 (C_7), 106.9 (C_4), 108.3 (C_1), 128.6 (C_6), 131.9 (C_5), 148.7 (C_3), 153.1 (C_2) and 190.2 (C_8) ppm.

Use this information to produce schematic diagrams of the HSQC and HMBC spectra, showing where all of the cross-peaks and diagonal peaks would be.

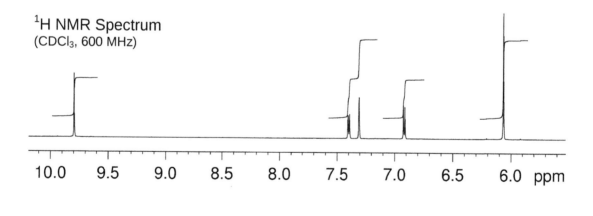

^1H NMR Spectrum
($CDCl_3$, 600 MHz)

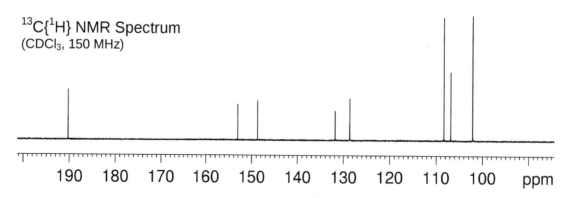

$^{13}C\{^1H\}$ NMR Spectrum
($CDCl_3$, 150 MHz)

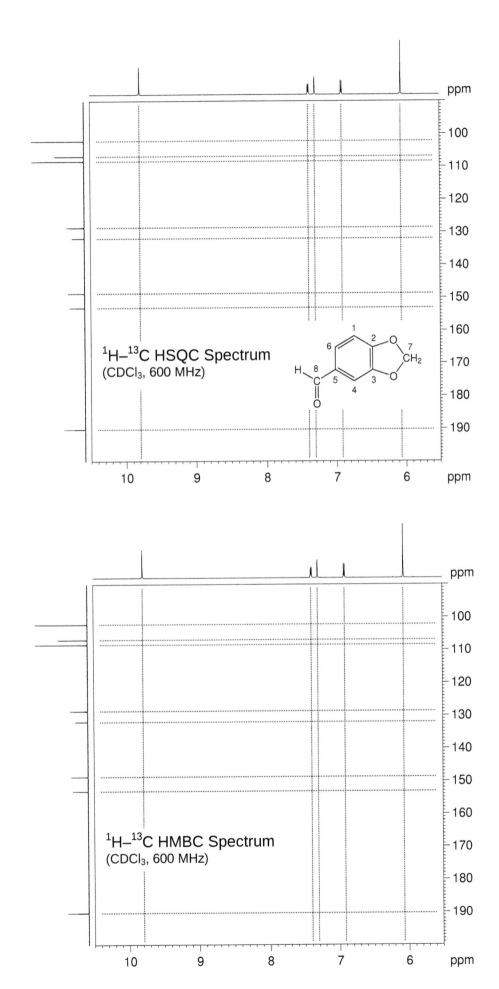

^{1}H–^{13}C HSQC Spectrum
(CDCl$_3$, 600 MHz)

^{1}H–^{13}C HMBC Spectrum
(CDCl$_3$, 600 MHz)

157

Problem 30

Identify the following compound.

Molecular Formula: $C_{13}H_{16}O_2$

IR: 1720 cm^{-1}

^{1}H NMR Spectrum
(CDCl$_3$, 400 MHz)

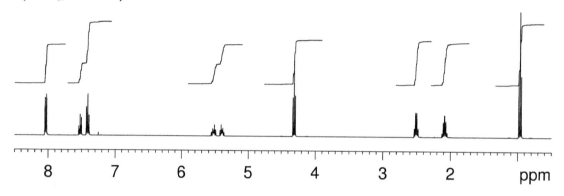

^{1}H NMR Expansion
(CDCl$_3$, 400 MHz)

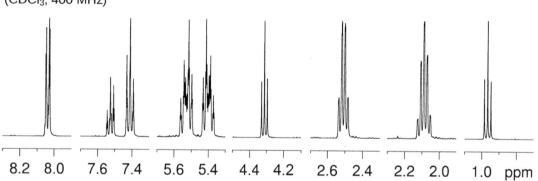

$^{13}C\{^1H\}$ NMR Spectrum
(CDCl$_3$, 100 MHz)

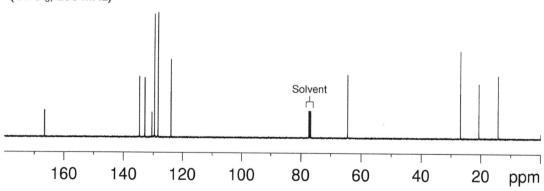

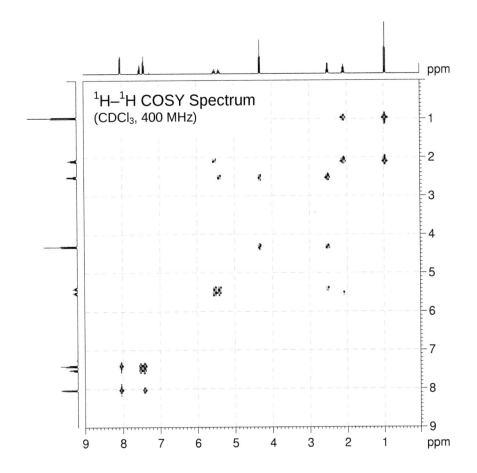

¹H–¹H COSY Spectrum
(CDCl₃, 400 MHz)

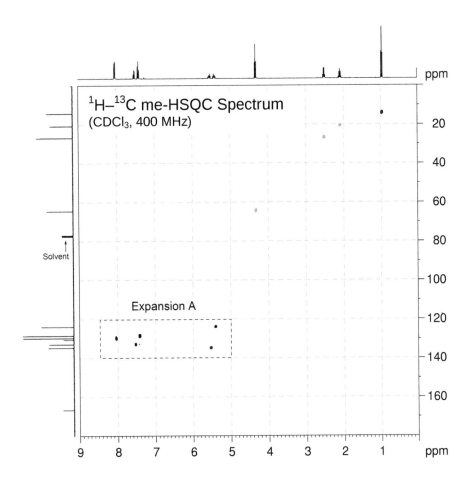

¹H–¹³C me-HSQC Spectrum
(CDCl₃, 400 MHz)

Solvent

Expansion A

159

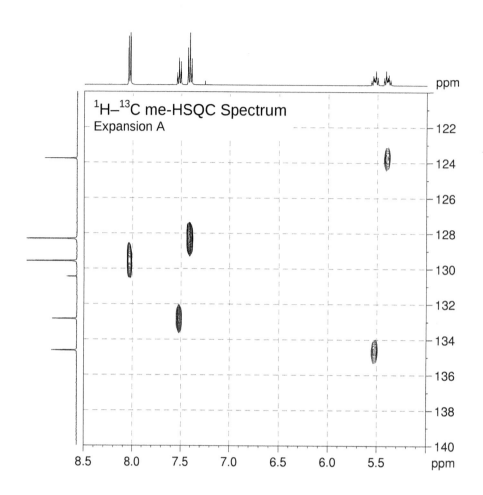

$^1H–^{13}C$ me-HSQC Spectrum
Expansion A

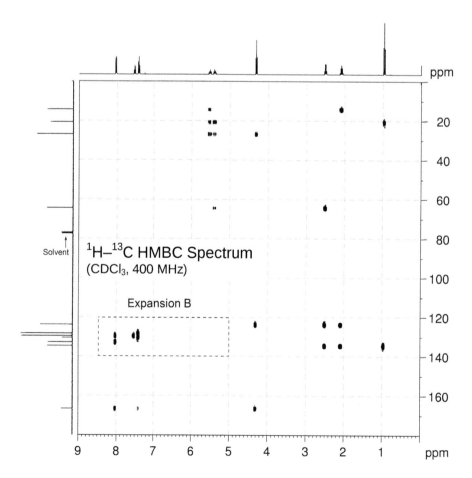

$^1H–^{13}C$ HMBC Spectrum
(CDCl$_3$, 400 MHz)

Expansion B

Solvent

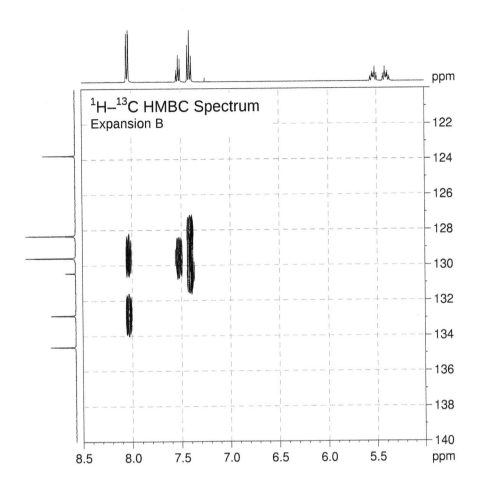

¹H–¹³C HMBC Spectrum
Expansion B

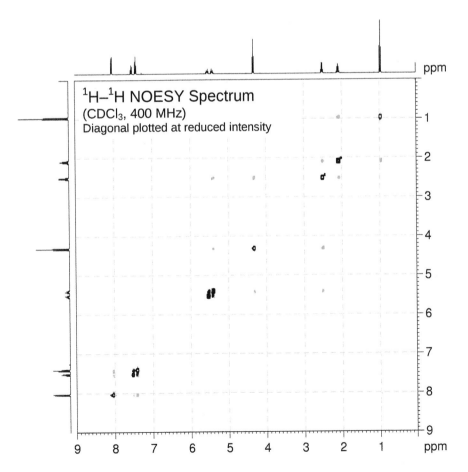

¹H–¹H NOESY Spectrum
(CDCl₃, 400 MHz)
Diagonal plotted at reduced intensity

161

Problem 31

Identify the following compound.
Molecular Formula: C$_9$H$_{14}$O
IR: 1693 cm^{-1}

^{1}H NMR Spectrum
(DMSO-d_6, 400 MHz)

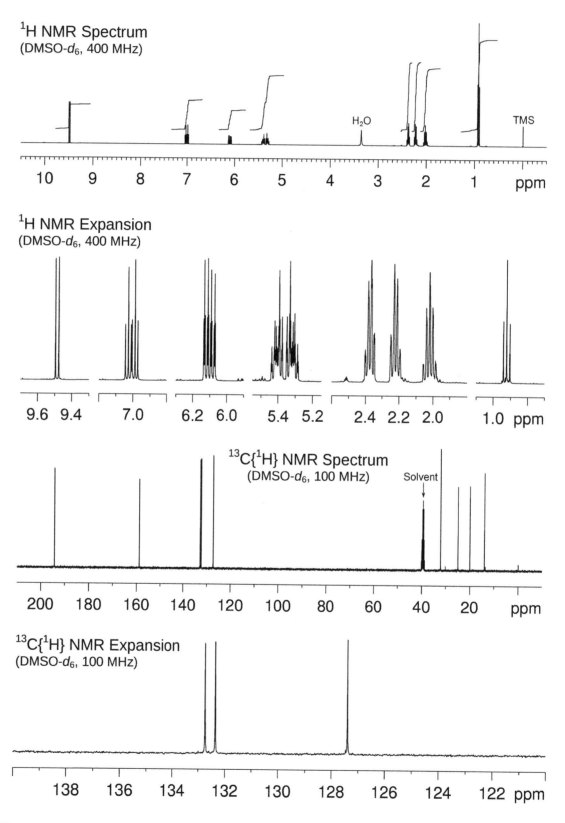

^{1}H NMR Expansion
(DMSO-d_6, 400 MHz)

^{13}C{^{1}H} NMR Spectrum
(DMSO-d_6, 100 MHz)

Solvent

^{13}C{^{1}H} NMR Expansion
(DMSO-d_6, 100 MHz)

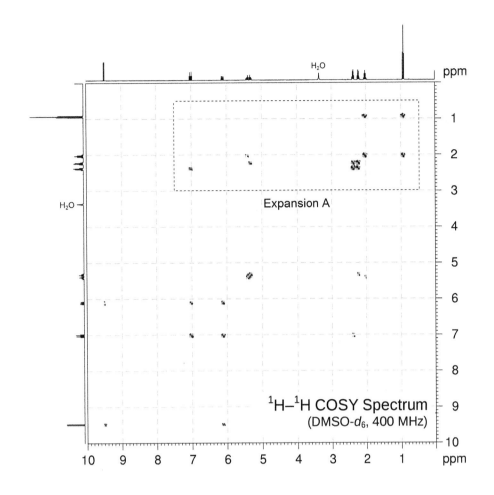

Expansion A

^{1}H–^{1}H COSY Spectrum
(DMSO-d_6, 400 MHz)

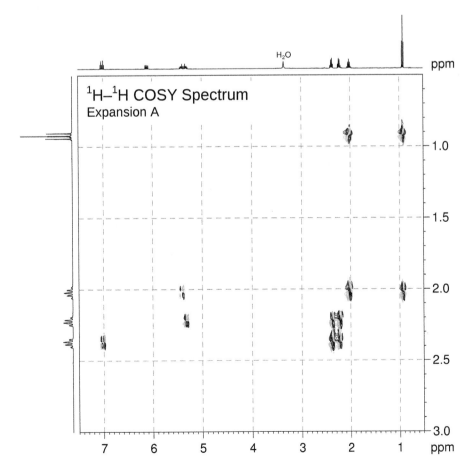

^{1}H–^{1}H COSY Spectrum
Expansion A

163

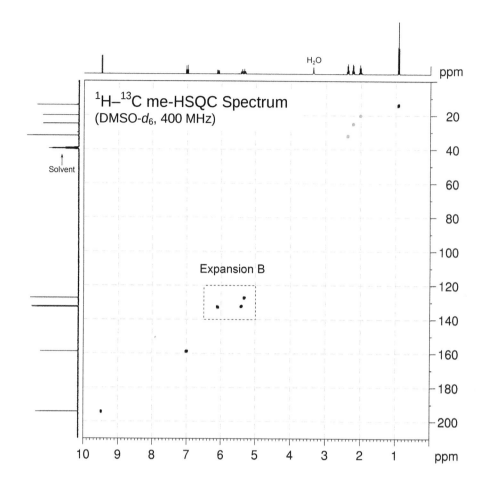

^{1}H–^{13}C me-HSQC Spectrum
(DMSO-d_6, 400 MHz)

Solvent

Expansion B

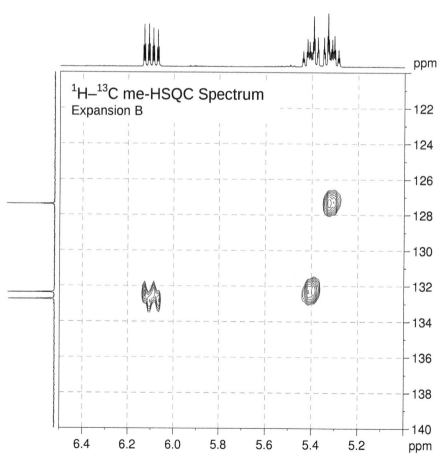

^{1}H–^{13}C me-HSQC Spectrum
Expansion B

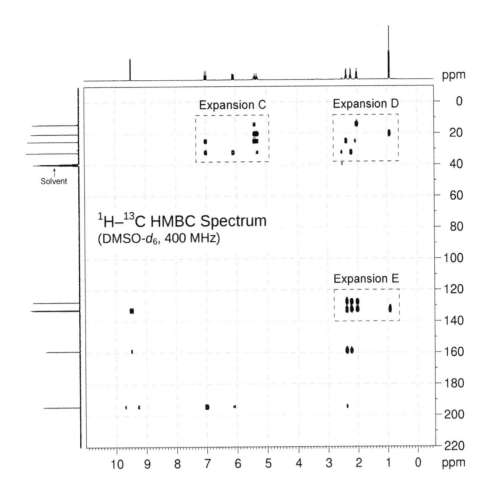

^{1}H–^{13}C HMBC Spectrum
(DMSO-d_6, 400 MHz)

Expansion C

Expansion D

Expansion E

Solvent

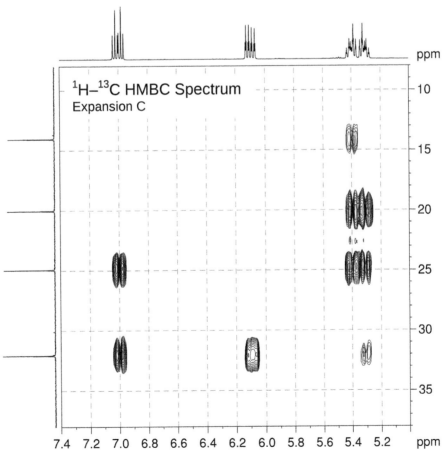

^{1}H–^{13}C HMBC Spectrum
Expansion C

165

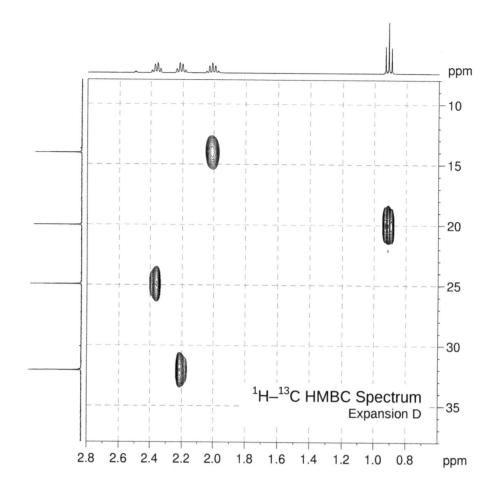

^{1}H–^{13}C HMBC Spectrum
Expansion D

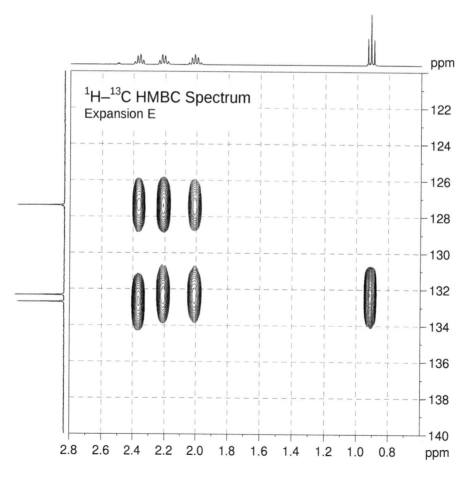

^{1}H–^{13}C HMBC Spectrum
Expansion E

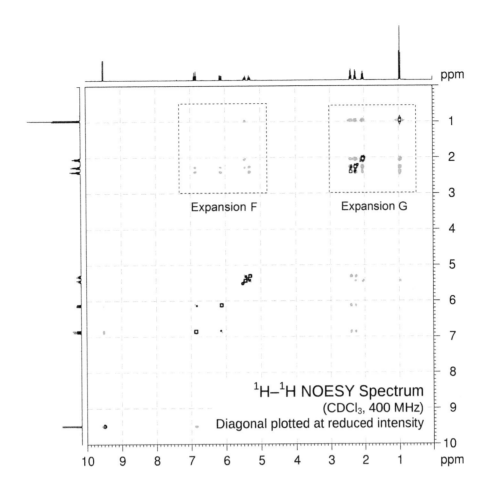

Expansion F

Expansion G

^{1}H–^{1}H NOESY Spectrum
(CDCl$_3$, 400 MHz)
Diagonal plotted at reduced intensity

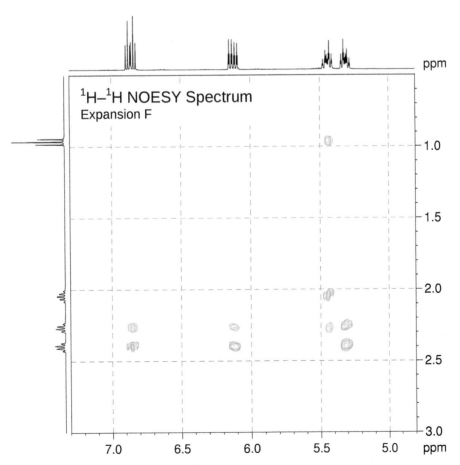

^{1}H–^{1}H NOESY Spectrum
Expansion F

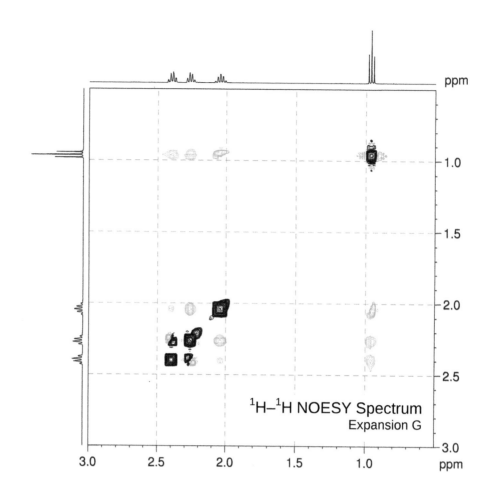

$^1H–^1H$ NOESY Spectrum
Expansion G

Problem 32

Identify the following compound.

Molecular Formula: $C_6H_{10}O_2$

IR Spectrum: 1649 (w), 1097 cm^{-1}

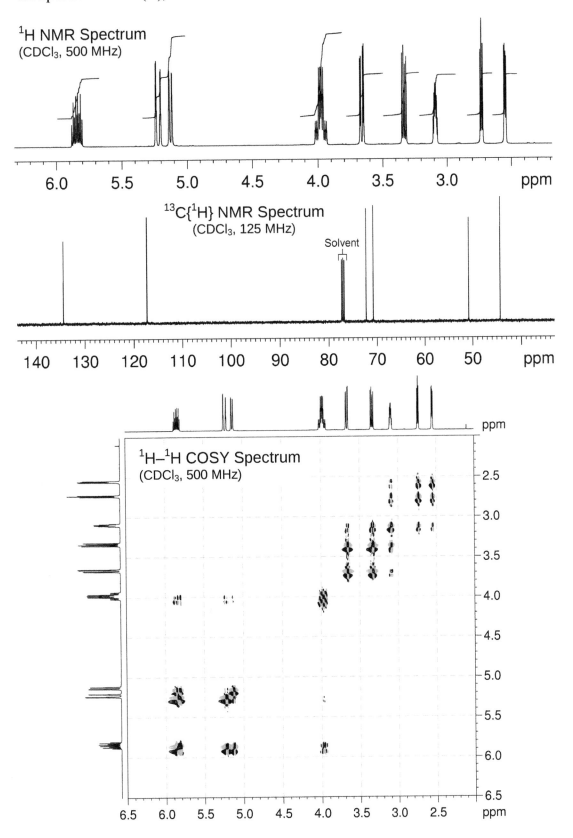

^{1}H NMR Spectrum
(CDCl$_3$, 500 MHz)

^{13}C{^{1}H} NMR Spectrum
(CDCl$_3$, 125 MHz)

Solvent

^{1}H–^{1}H COSY Spectrum
(CDCl$_3$, 500 MHz)

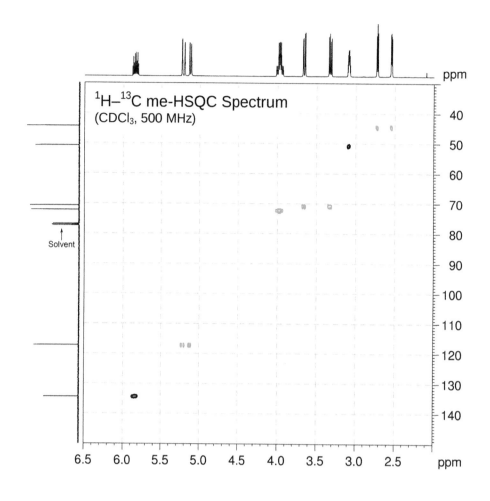

^{1}H–^{13}C me-HSQC Spectrum
(CDCl$_3$, 500 MHz)

Solvent

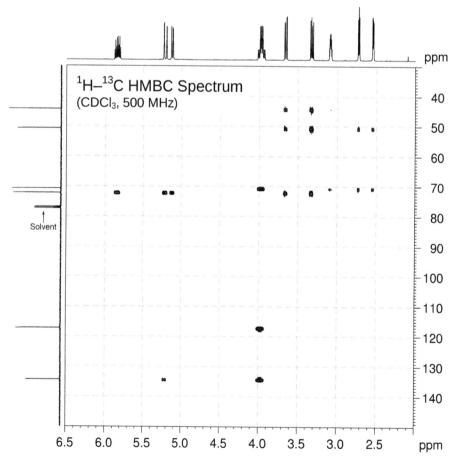

^{1}H–^{13}C HMBC Spectrum
(CDCl$_3$, 500 MHz)

Solvent

Problem 33

Identify the following compound.

Molecular Formula: $C_6H_{10}O_2$

IR: 1712 cm^{-1}

^{1}H NMR Spectrum
(CDCl$_3$, 500 MHz)

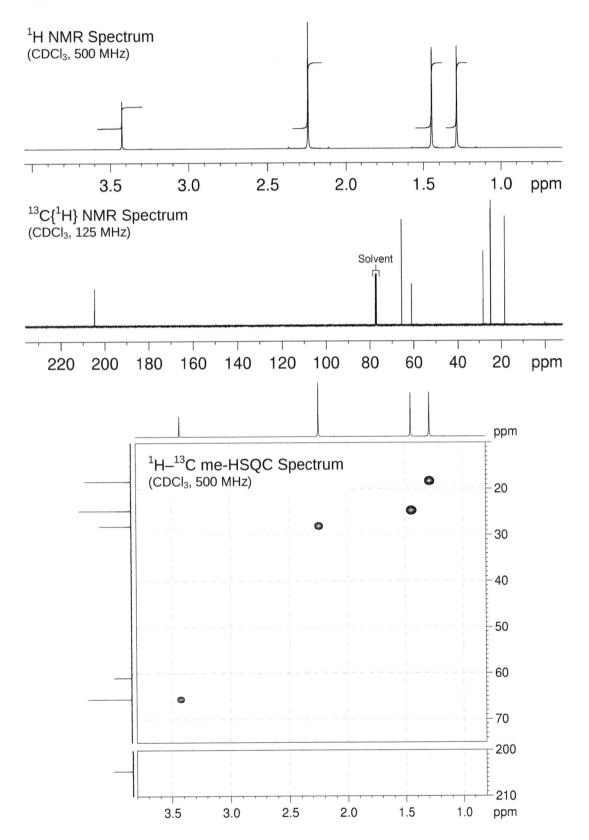

^{13}C{^{1}H} NMR Spectrum
(CDCl$_3$, 125 MHz)

Solvent

^{1}H–^{13}C me-HSQC Spectrum
(CDCl$_3$, 500 MHz)

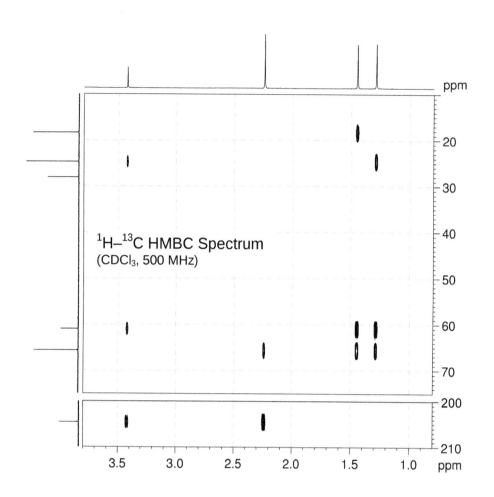

¹H–¹³C HMBC Spectrum
(CDCl₃, 500 MHz)

Problem 34

Given below are six compounds which are isomers of $C_5H_{11}NO_2S$. The ^1H and $^{13}C\{^1H\}$ NMR spectra of one of the isomers are given below. The ^1H–^{13}C me-HSQC and ^1H–^{13}C HMBC spectra are given on the following page. To which of these compounds do the spectra belong?

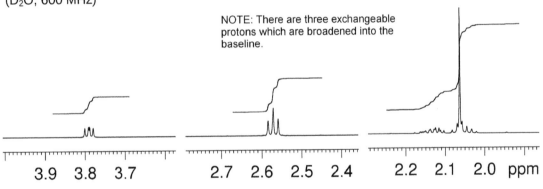

A   B   C

D   E   F

^1H NMR Expansions
(D₂O, 600 MHz)

NOTE: There are three exchangeable protons which are broadened into the baseline.

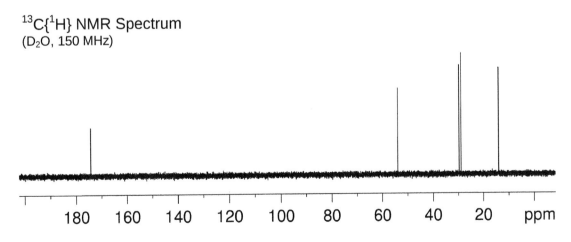

3.9 3.8 3.7

2.7 2.6 2.5 2.4

2.2 2.1 2.0 ppm

$^{13}C\{^1H\}$ NMR Spectrum
(D₂O, 150 MHz)

180   160   140   120   100   80   60   40   20   ppm

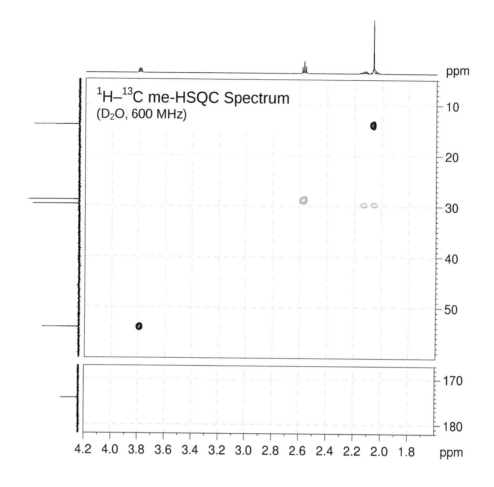

¹H–¹³C me-HSQC Spectrum
(D₂O, 600 MHz)

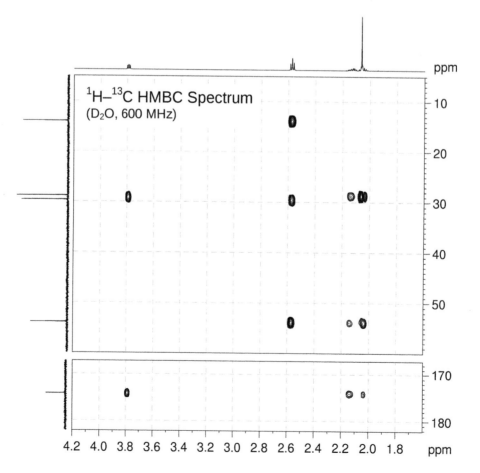

¹H–¹³C HMBC Spectrum
(D₂O, 600 MHz)

Problem 35

Identify the following compound.

Molecular Formula: $C_8H_{15}NO_3$

IR: 3333, 1702, 1626 cm^{-1}

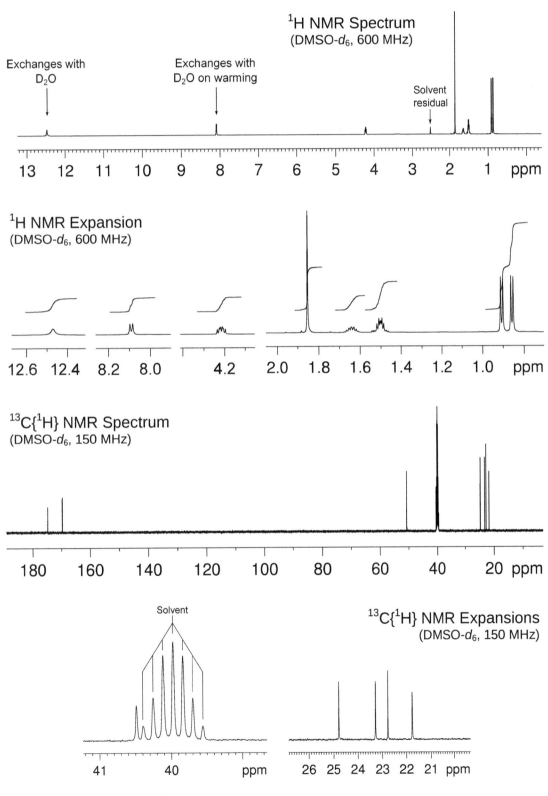

^{1}H NMR Spectrum
(DMSO-d_6, 600 MHz)

Exchanges with D$_2$O

Exchanges with D$_2$O on warming

Solvent residual

^{1}H NMR Expansion
(DMSO-d_6, 600 MHz)

^{13}C{^{1}H} NMR Spectrum
(DMSO-d_6, 150 MHz)

Solvent

^{13}C{^{1}H} NMR Expansions
(DMSO-d_6, 150 MHz)

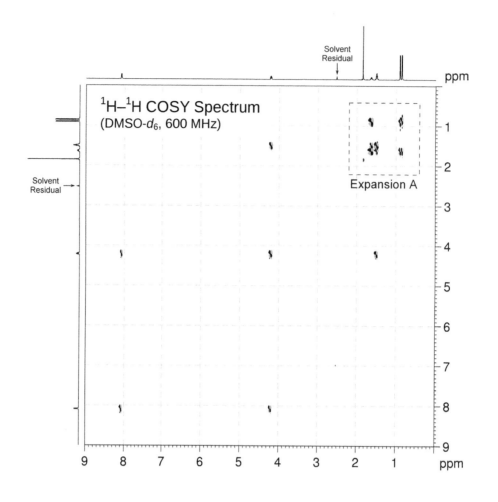

^1H–^1H COSY Spectrum
(DMSO-d_6, 600 MHz)

Solvent
Residual

Solvent
Residual

Expansion A

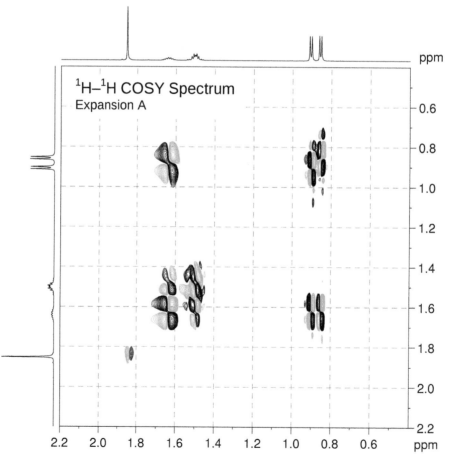

^1H–^1H COSY Spectrum
Expansion A

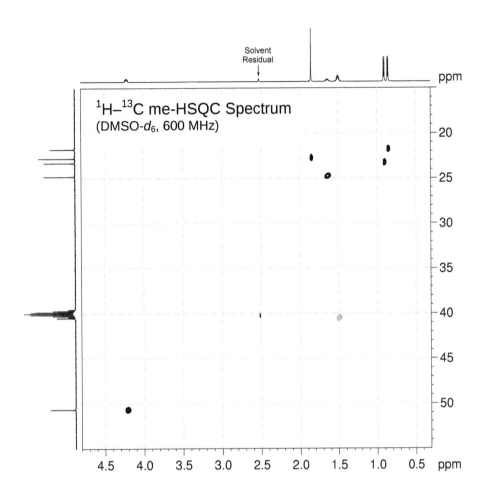

¹H–¹³C me-HSQC Spectrum
(DMSO-d_6, 600 MHz)

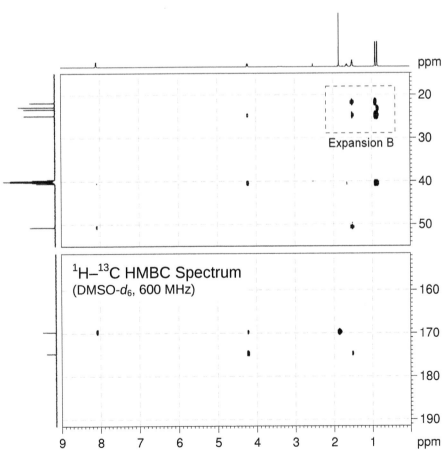

¹H–¹³C HMBC Spectrum
(DMSO-d_6, 600 MHz)

Expansion B

Solvent
Residual

177

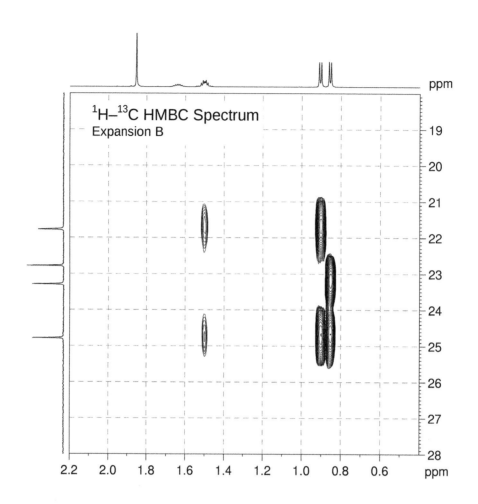

Problem 36

Identify the following compound.

Molecular Formula: $C_{10}H_{20}O_2$

IR: 1739 cm^{-1}

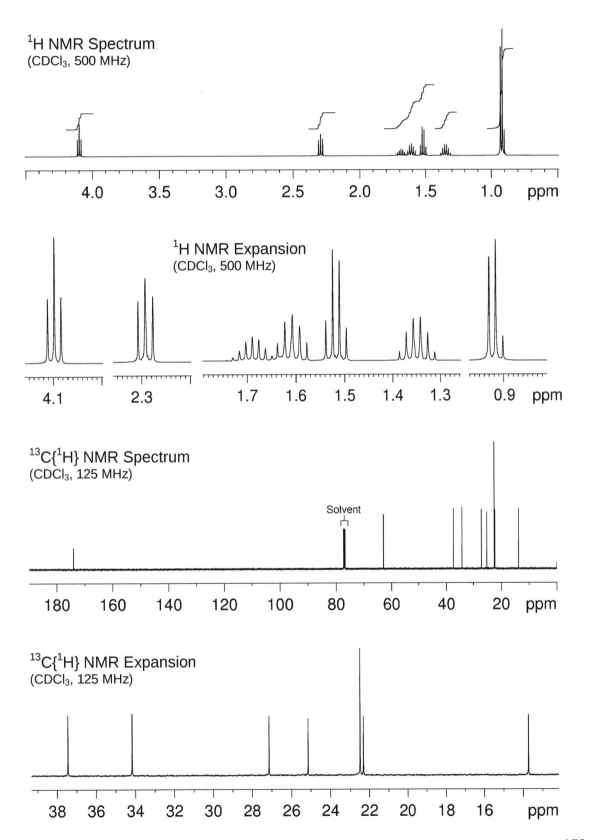

^{1}H NMR Spectrum
(CDCl$_3$, 500 MHz)

^{1}H NMR Expansion
(CDCl$_3$, 500 MHz)

^{13}C{^{1}H} NMR Spectrum
(CDCl$_3$, 125 MHz)

Solvent

^{13}C{^{1}H} NMR Expansion
(CDCl$_3$, 125 MHz)

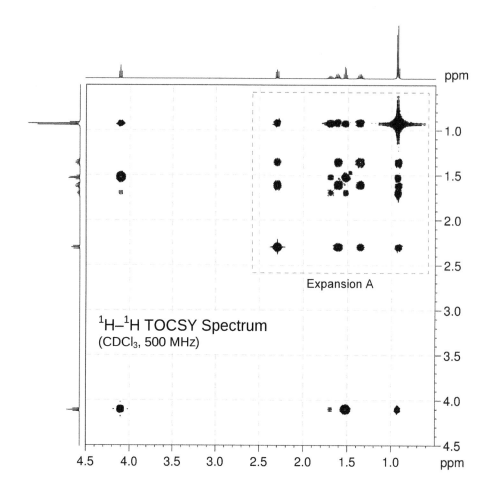

^{1}H–^{1}H TOCSY Spectrum
(CDCl$_3$, 500 MHz)

Expansion A

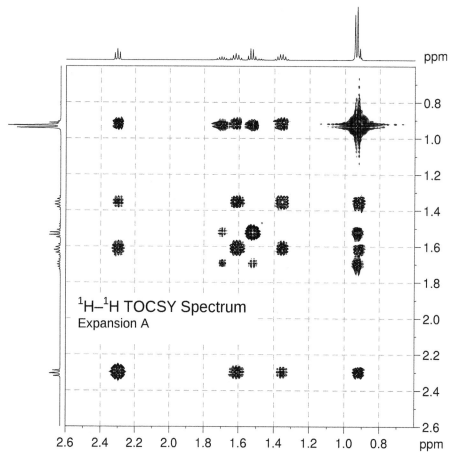

^{1}H–^{1}H TOCSY Spectrum
Expansion A

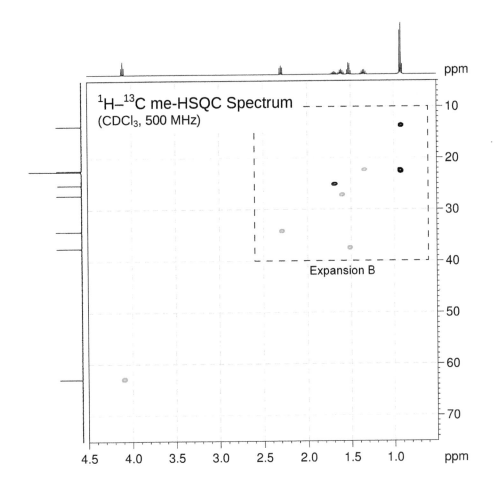

¹H–¹³C me-HSQC Spectrum
(CDCl₃, 500 MHz)

Expansion B

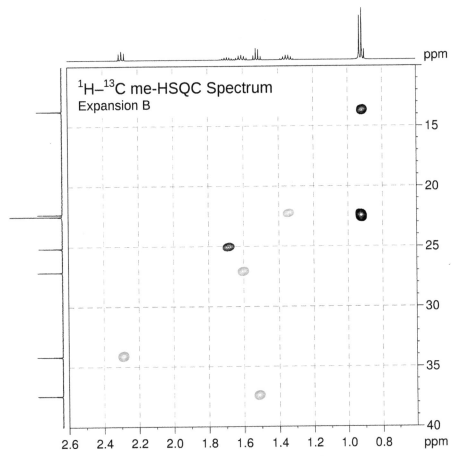

¹H–¹³C me-HSQC Spectrum
Expansion B

Problem 37

The ^{1}H and ^{13}C{^{1}H} NMR spectra of (E)-4-methyl-4'-nitrostilbene ($C_{15}H_{13}NO_2$) recorded in acetone-d_6 solution at 298 K and 500 MHz are given below.

The ^{1}H NMR spectrum has signals at δ 2.34, 7.23, 7.33, 7.48, 7.56, 7.83 and 8.22 ppm.

The ^{13}C{^{1}H} NMR spectrum has signals at δ 21.3, 124.8, 126.2, 127.9, 128.0, 130.3, 134.1, 134.8, 139.5, 145.3 and 147.5 ppm.

The ^{1}H–^{1}H COSY, ^{1}H–^{13}C me-HSQC and ^{1}H–^{13}C HMBC spectra are given on the following pages. Use these spectra to assign the ^{1}H and ^{13}C{^{1}H} resonances for this compound.

Proton	Chemical Shift (ppm)	Carbon	Chemical Shift (ppm)
H_1		C_1	
		C_2	
H_3		C_3	
H_4		C_4	
		C_5	
H_6		C_6	
H_7		C_7	
		C_8	
H_9		C_9	
H_{10}		C_{10}	
		C_{11}	

^{1}H NMR Spectrum
(Acetone-d_6, 500 MHz)

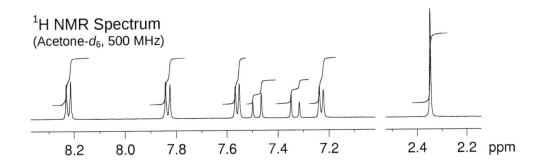

^{13}C{^{1}H} NMR Spectrum
(Acetone-d_6, 125 MHz)

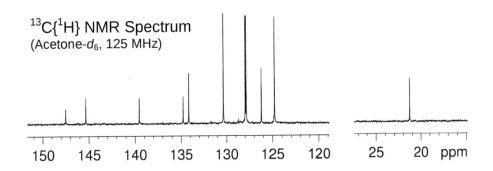

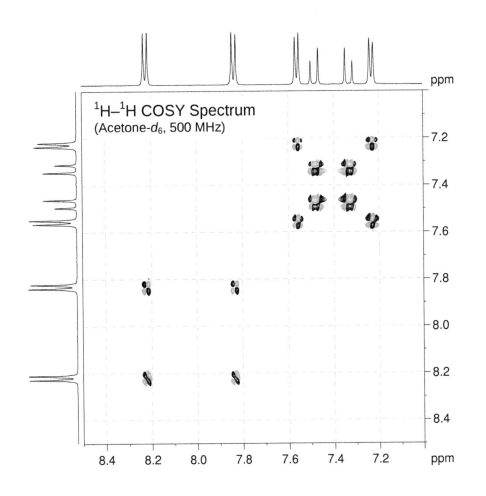

^{1}H–^{1}H COSY Spectrum
(Acetone-d_6, 500 MHz)

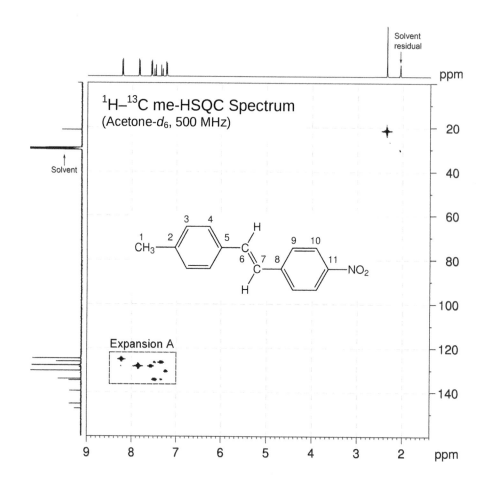

^{1}H–^{13}C me-HSQC Spectrum
(Acetone-d_6, 500 MHz)

Solvent residual

Solvent

Expansion A

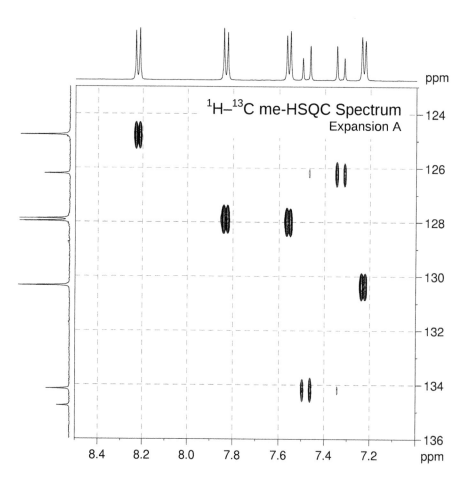

^{1}H–^{13}C me-HSQC Spectrum
Expansion A

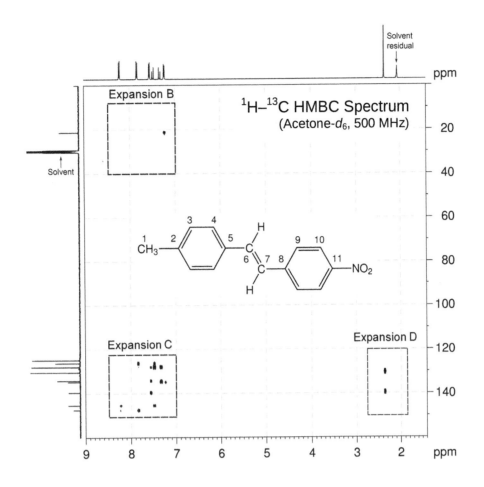

$^1H–^{13}C$ HMBC Spectrum
(Acetone-d_6, 500 MHz)

Expansion B

Expansion C

Expansion D

Solvent residual

Solvent

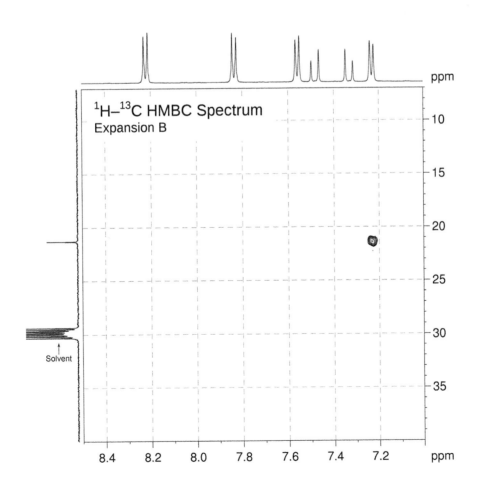

$^1H–^{13}C$ HMBC Spectrum
Expansion B

Solvent

185

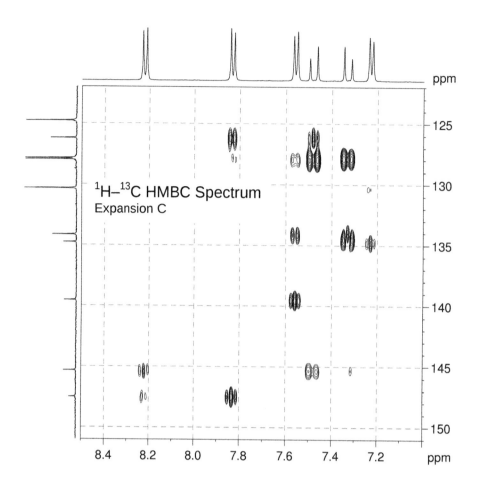

¹H–¹³C HMBC Spectrum
Expansion C

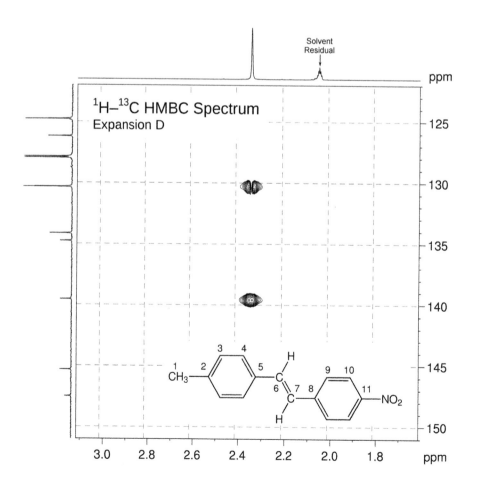

¹H–¹³C HMBC Spectrum
Expansion D

Solvent Residual

186

Problem 38

Identify the following compound.
Molecular Formula: $C_{11}H_{16}O$
IR: 3450 (br) cm^{-1}

^{1}H NMR Spectrum
(CDCl$_3$, 600 MHz)

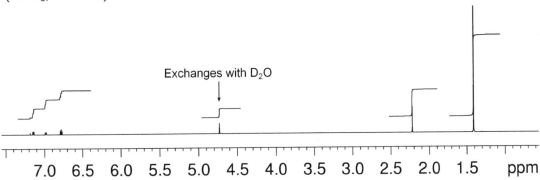

Exchanges with D$_2$O

^{1}H NMR Expansion Spectrum
(CDCl$_3$, 600 MHz)

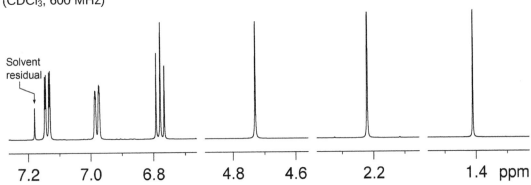

Solvent residual

^{13}C{^{1}H} NMR Spectrum
(CDCl$_3$, 150 MHz)

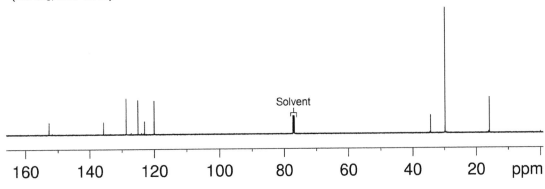

Solvent

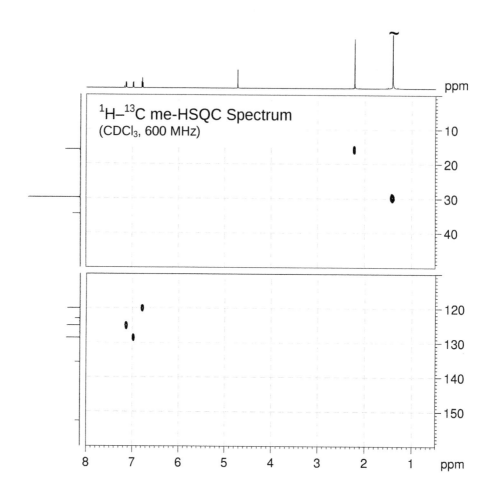

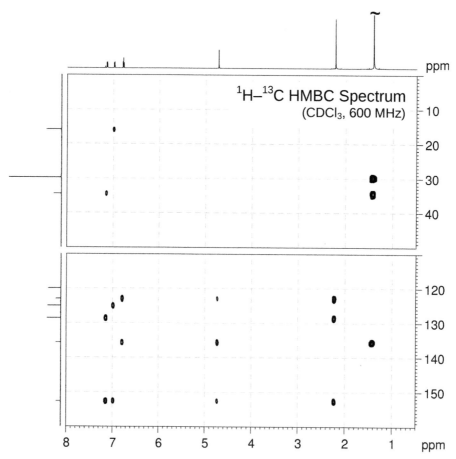

Problem 39

Identify the following compound.

Molecular Formula: $C_{10}H_{12}O$

IR: 3600 (br), 1638, 1594, 1469 (s) cm^{-1}

^{1}H NMR Spectrum
(CDCl$_3$, 500 MHz)

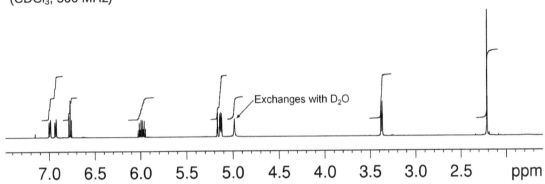

Exchanges with D$_2$O

^{1}H NMR Expansion Spectrum
(CDCl$_3$, 500 MHz)

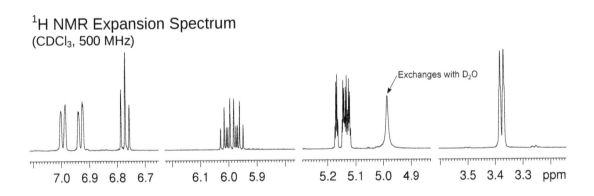

Exchanges with D$_2$O

^{13}C{^{1}H} NMR Spectrum
(CDCl$_3$, 125 MHz)

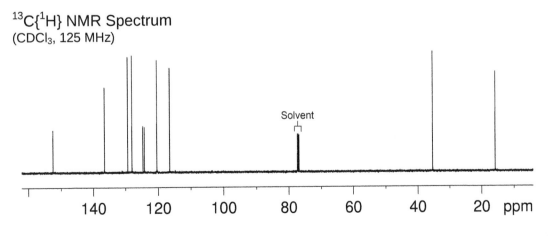

Solvent

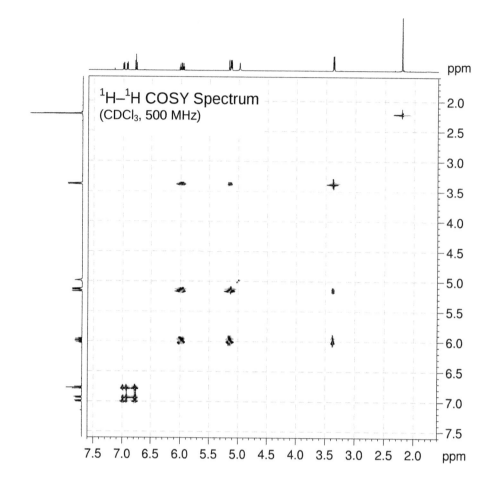

^{1}H–^{1}H COSY Spectrum
(CDCl$_3$, 500 MHz)

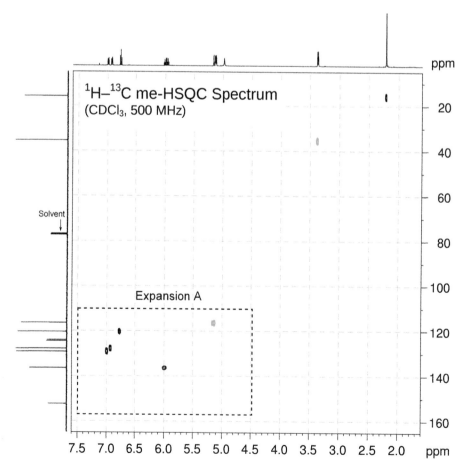

^{1}H–^{13}C me-HSQC Spectrum
(CDCl$_3$, 500 MHz)

Solvent

Expansion A

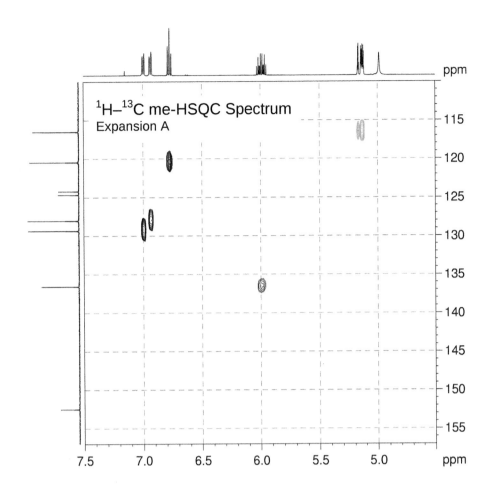

^{1}H–^{13}C me-HSQC Spectrum
Expansion A

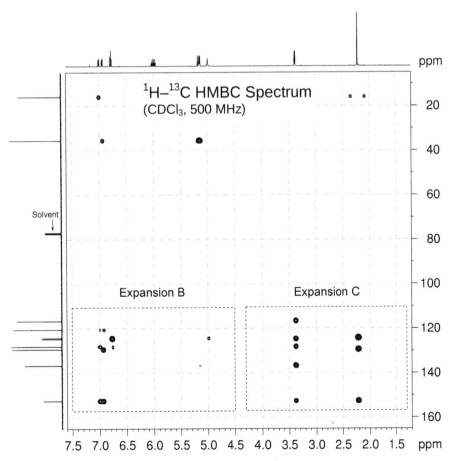

^{1}H–^{13}C HMBC Spectrum
(CDCl$_3$, 500 MHz)

Solvent

Expansion B

Expansion C

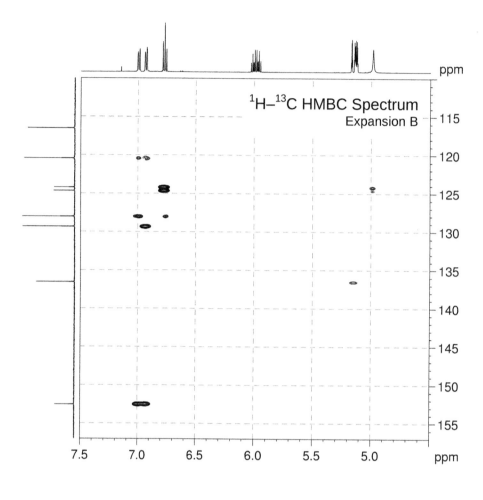

¹H–¹³C HMBC Spectrum
Expansion B

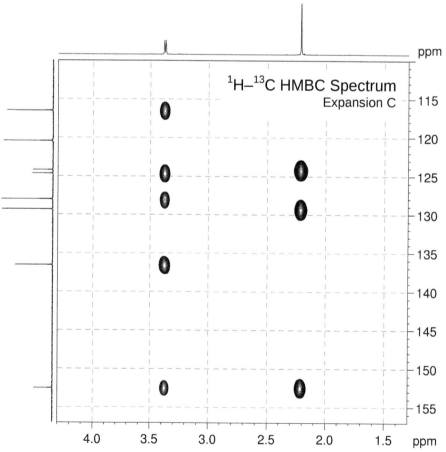

¹H–¹³C HMBC Spectrum
Expansion C

Problem 40

Identify the following compound.
Molecular Formula: $C_8H_8O_3$
IR: 3022 (br w), 1675, 1636 cm^{-1}

^{1}H NMR Spectrum
(CDCl$_3$, 600 MHz)

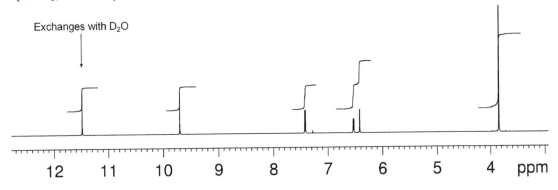

^{1}H NMR Expansion Spectrum
(CDCl$_3$, 600 MHz)

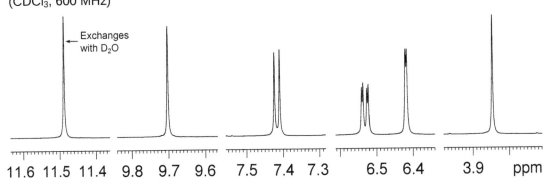

^{13}C{^{1}H} NMR Spectrum
(CDCl$_3$, 150 MHz)

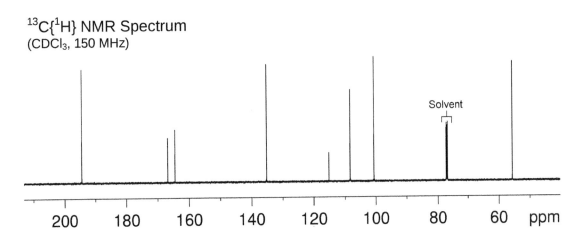

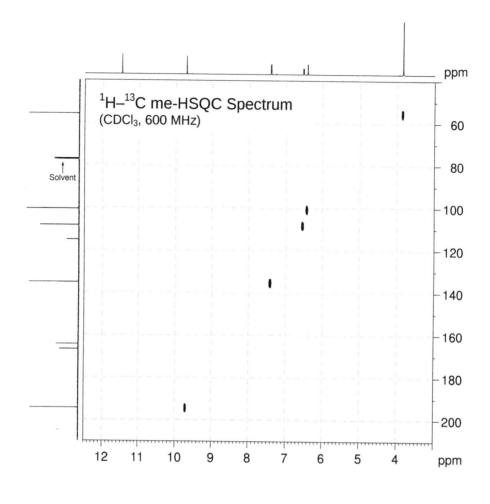

^1H–^{13}C me-HSQC Spectrum
(CDCl$_3$, 600 MHz)

Solvent

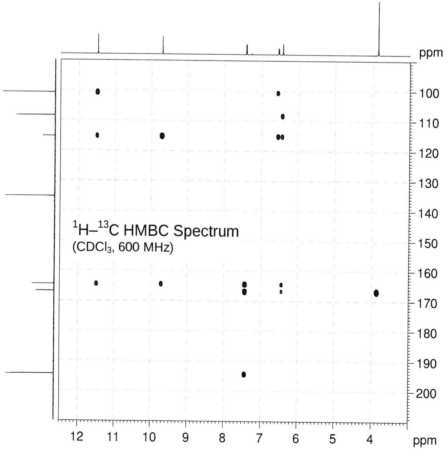

^1H–^{13}C HMBC Spectrum
(CDCl$_3$, 600 MHz)

Problem 41

Identify the following compound.

Molecular Formula: $C_9H_{10}O_2$

IR: 1644, 1618 cm^{-1}

^{1}H NMR Spectrum
(CDCl$_3$, 600 MHz)

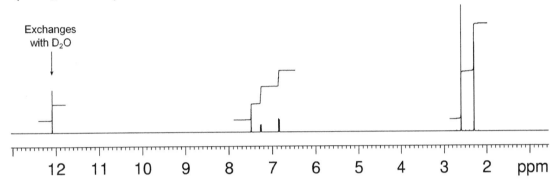

^{1}H NMR Expansion Spectrum
(CDCl$_3$, 600 MHz)

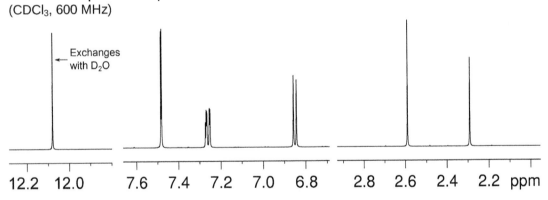

^{13}C{^{1}H} NMR Spectrum
(CDCl$_3$, 150 MHz)

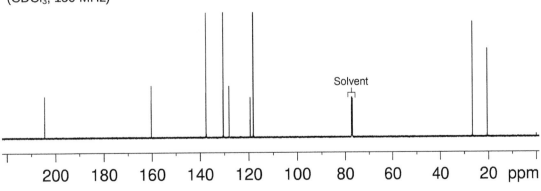

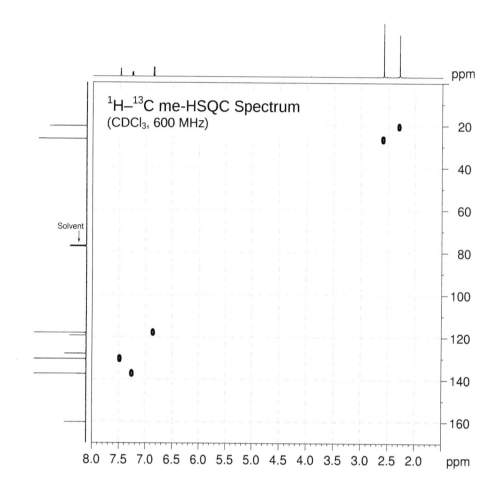

^1H–^{13}C me-HSQC Spectrum
(CDCl$_3$, 600 MHz)

Solvent

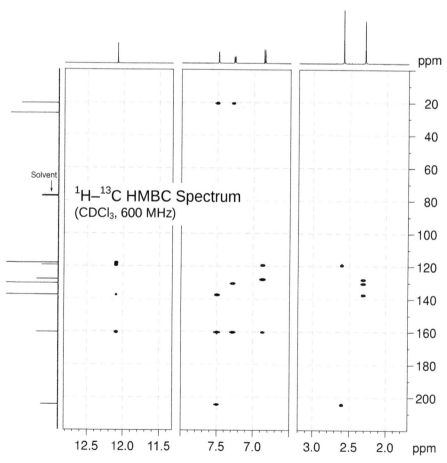

^1H–^{13}C HMBC Spectrum
(CDCl$_3$, 600 MHz)

Solvent

Problem 42

Identify the following compound.
Molecular Formula: $C_9H_9FO_2$
IR: 1681 cm^{-1}

^{1}H NMR Spectrum
(DMSO-d_6, 400 MHz)

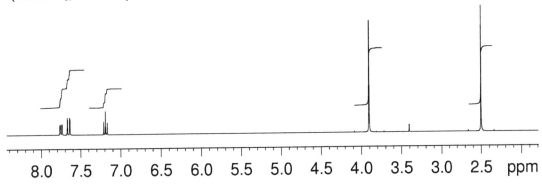

^{1}H NMR Expansions
(DMSO-d_6, 400 MHz)

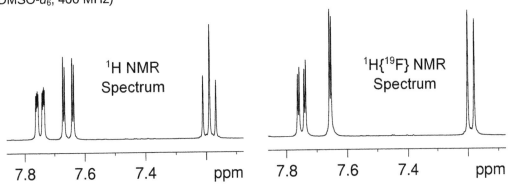

^{1}H NMR
Spectrum

^{1}H{^{19}F} NMR
Spectrum

^{19}F{^{1}H} NMR Spectrum
(DMSO-d_6, 376 MHz)

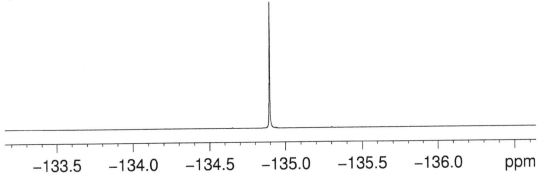

$^{13}\text{C}\{^1\text{H}\}$ NMR Spectrum
(DMSO-d_6, 100 MHz)

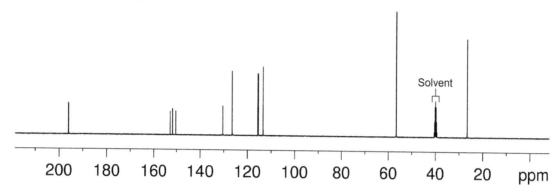

$^{13}\text{C}\{^1\text{H}\}$ NMR Expansions
(DMSO-d_6, 100 MHz)

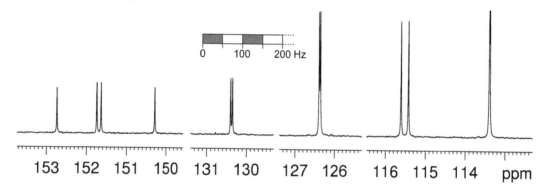

$^{13}\text{C}\{^1\text{H},^{19}\text{F}\}$ NMR Expansions
(DMSO-d_6, 100 MHz)

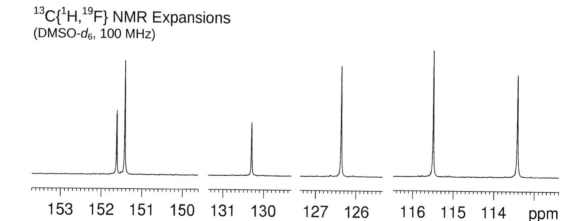

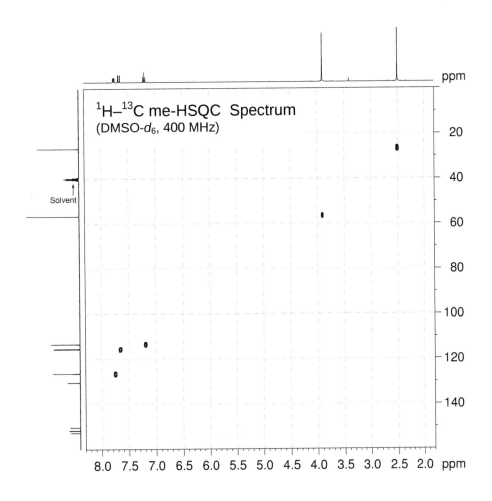

^{1}H–^{13}C me-HSQC Spectrum
(DMSO-d_6, 400 MHz)

Solvent

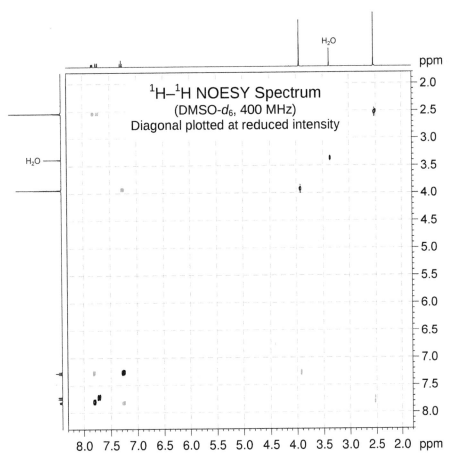

^{1}H–^{1}H NOESY Spectrum
(DMSO-d_6, 400 MHz)
Diagonal plotted at reduced intensity

H_2O

Problem 43

Identify the following compound.
Molecular Formula: $C_{10}H_{10}O_4$

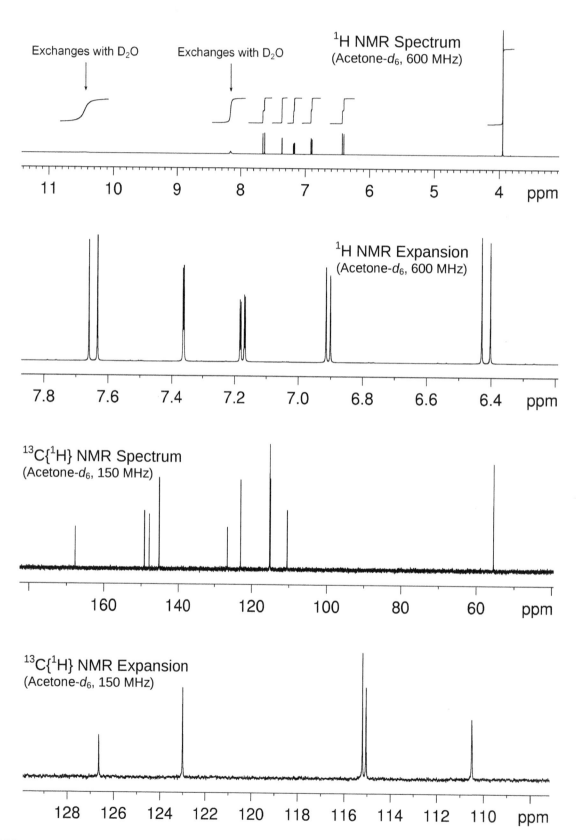

Exchanges with D₂O

Exchanges with D₂O

¹H NMR Spectrum
(Acetone-d_6, 600 MHz)

¹H NMR Expansion
(Acetone-d_6, 600 MHz)

¹³C{¹H} NMR Spectrum
(Acetone-d_6, 150 MHz)

¹³C{¹H} NMR Expansion
(Acetone-d_6, 150 MHz)

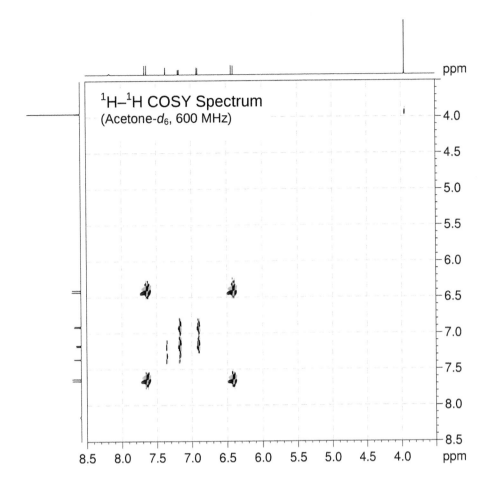

¹H–¹H COSY Spectrum
(Acetone-*d*₆, 600 MHz)

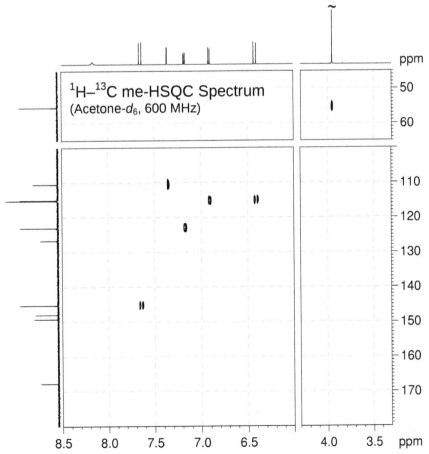

¹H–¹³C me-HSQC Spectrum
(Acetone-*d*₆, 600 MHz)

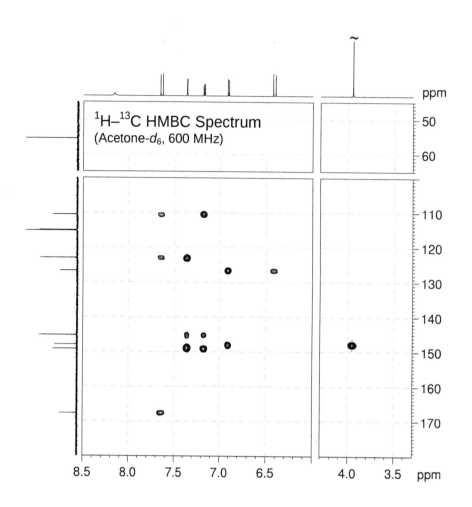

¹H–¹³C HMBC Spectrum
(Acetone-*d*₆, 600 MHz)

202

Problem 44

Identify the following compound.

Molecular Formula: $C_{13}H_{16}O_3$

IR: 3385 (br), 2973, 1685, 1636, 1583 cm^{-1}

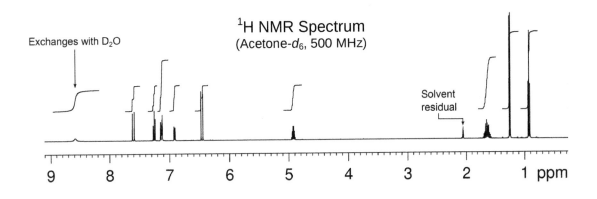

^{1}H NMR Spectrum
(Acetone-d_6, 500 MHz)

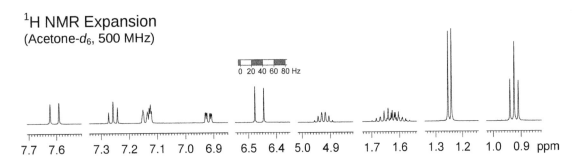

^{1}H NMR Expansion
(Acetone-d_6, 500 MHz)

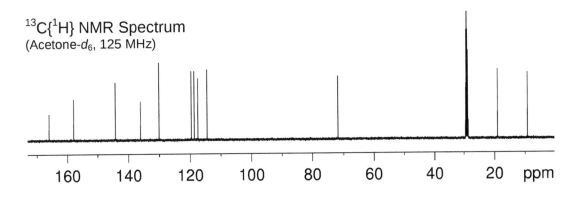

$^{13}C\{^1H\}$ NMR Spectrum
(Acetone-d_6, 125 MHz)

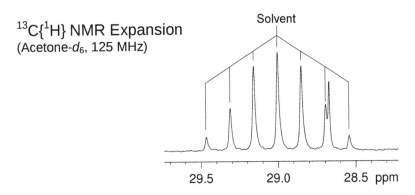

$^{13}C\{^1H\}$ NMR Expansion
(Acetone-d_6, 125 MHz)

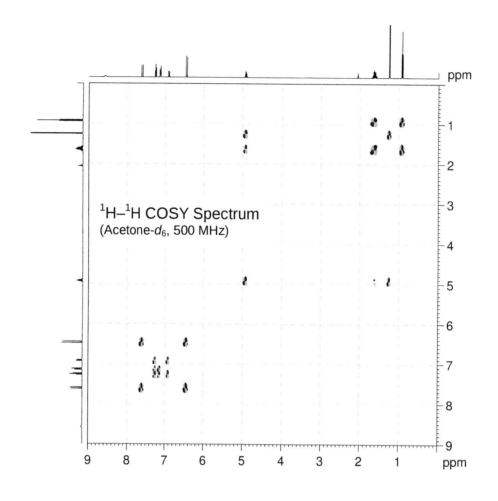

$^1H–^1H$ COSY Spectrum
(Acetone-d_6, 500 MHz)

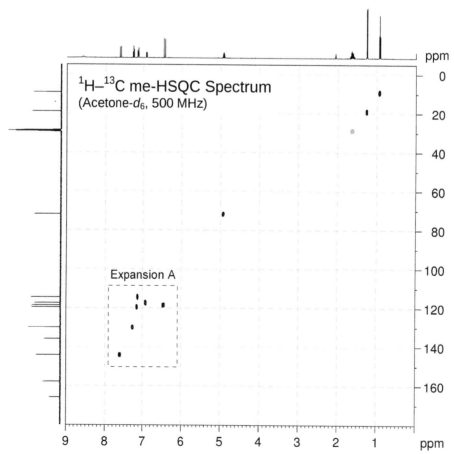

$^1H–^{13}C$ me-HSQC Spectrum
(Acetone-d_6, 500 MHz)

Expansion A

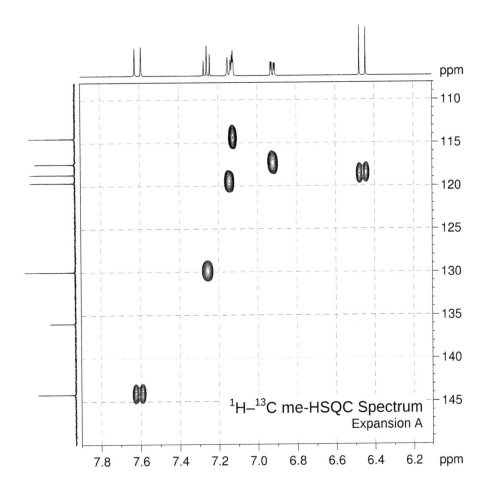

¹H–¹³C me-HSQC Spectrum
Expansion A

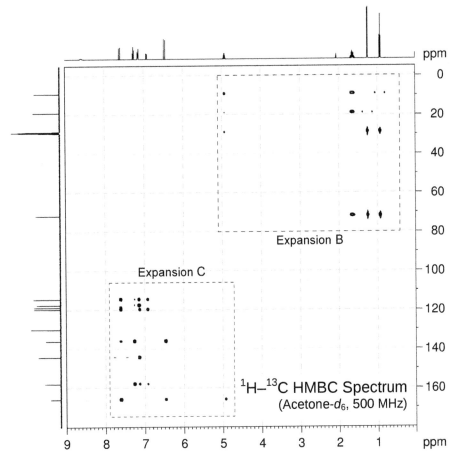

Expansion B

Expansion C

¹H–¹³C HMBC Spectrum
(Acetone-*d*₆, 500 MHz)

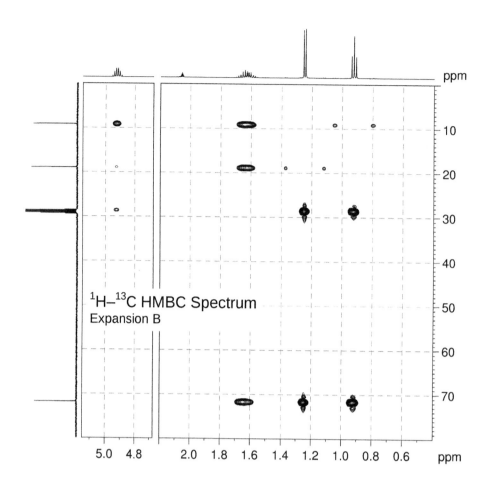

$^1H-^{13}C$ HMBC Spectrum
Expansion B

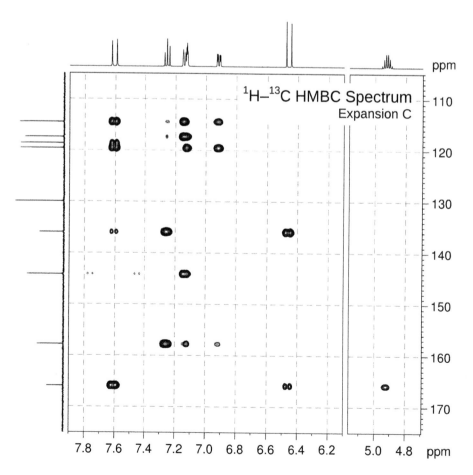

$^1H-^{13}C$ HMBC Spectrum
Expansion C

Problem 45

Identify the following compound. Draw a labelled structure, and use the ^1H–^1H COSY, ^1H–^{13}C me-HSQC and ^1H–^{13}C HMBC spectra to assign each of the ^1H and ^{13}C resonances to the appropriate carbons and hydrogens in the structure.

Molecular Formula: $C_{11}H_{12}O$

IR: 1676 cm^{-1}

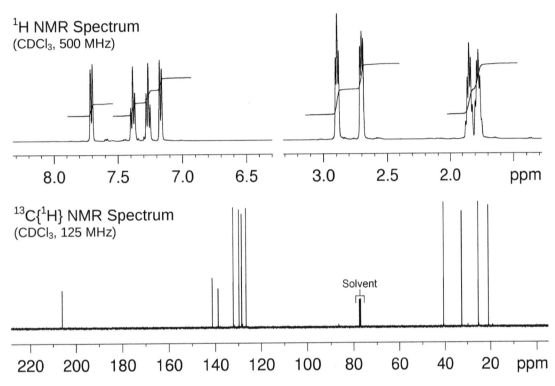

^1H NMR Spectrum (CDCl$_3$, 500 MHz)

$^{13}C\{^1H\}$ NMR Spectrum (CDCl$_3$, 125 MHz)

Proton	Chemical Shift (ppm)	Carbon	Chemical Shift (ppm)

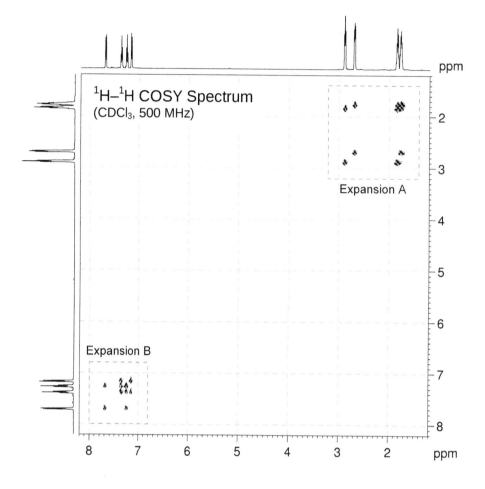

^{1}H–^{1}H COSY Spectrum
(CDCl$_3$, 500 MHz)

Expansion A

Expansion B

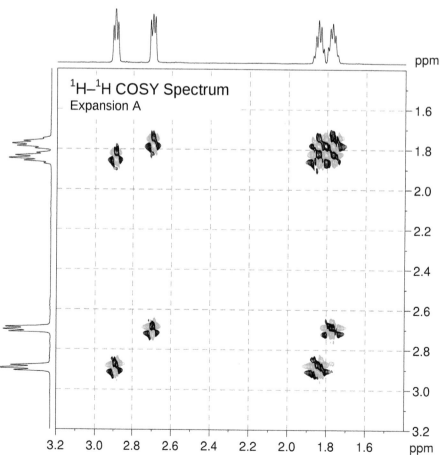

^{1}H–^{1}H COSY Spectrum
Expansion A

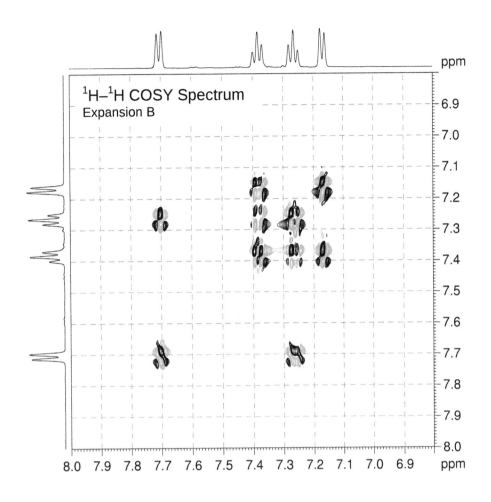

^{1}H–^{1}H COSY Spectrum
Expansion B

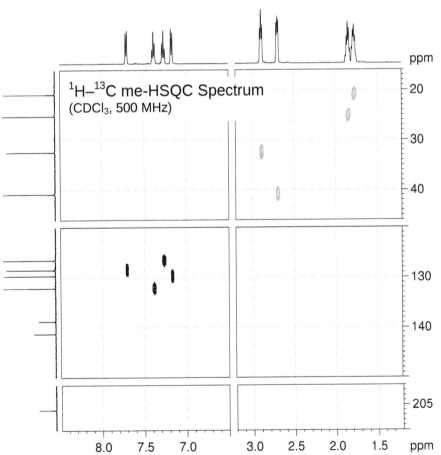

^{1}H–^{13}C me-HSQC Spectrum
(CDCl$_3$, 500 MHz)

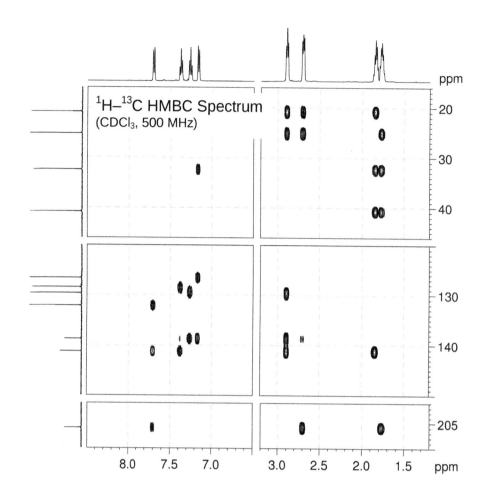

$^1H–^{13}C$ HMBC Spectrum
(CDCl$_3$, 500 MHz)

Structure:

210

Problem 46

Identify the following compound. Draw a labelled structure, and use the $^1H-^{13}C$ me-HSQC, $^1H-^1H$ COSY and $^1H-^{31}P$ HMBC spectra to assign each of the ^1H and ^{13}C resonances to the appropriate carbons and hydrogens in the structure.

Molecular Formula: $C_5H_{12}BrO_3P$

IR: 3464, 2955, 1232, 1031 cm^{-1}

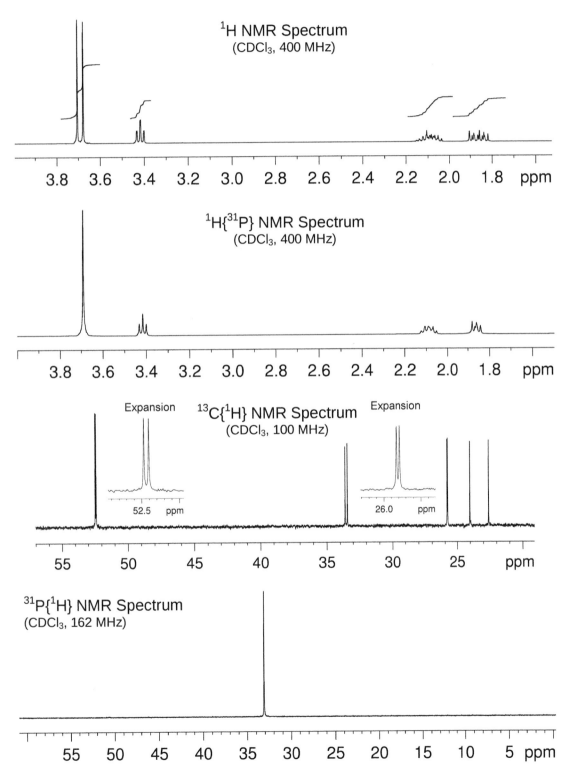

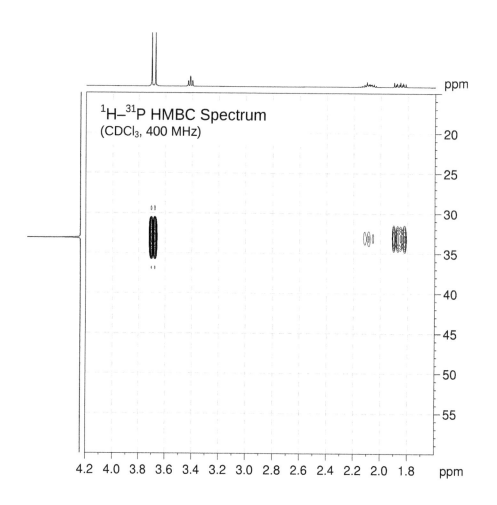

^{1}H–^{31}P HMBC Spectrum
(CDCl$_3$, 400 MHz)

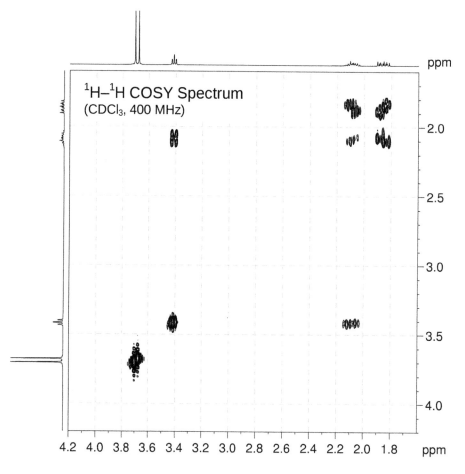

^{1}H–^{1}H COSY Spectrum
(CDCl$_3$, 400 MHz)

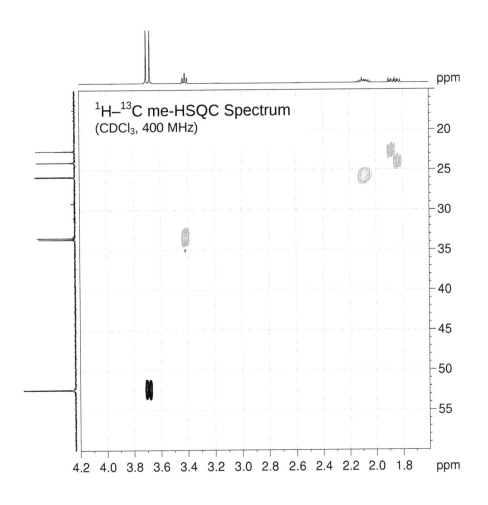

Structure:

Proton	Chemical Shift (ppm)	Carbon	Chemical Shift (ppm)

Problem 47

The ^{1}H and ^{13}C{^{1}H} NMR spectra of caffeine ($C_8H_{10}N_4O_2$) recorded in DMSO-d_6 solution at 298 K and 500 MHz are given below.

The ^{1}H NMR spectrum has signals at δ 3.16, 3.35, 3.84 and 7.96 ppm.

The ^{13}C{^{1}H} NMR spectrum has signals at δ 27.4, 29.3, 33.1, 106.5, 142.7, 148.0, 150.9 and 154.4 ppm.

The ^{15}N NMR spectrum has signals at δ 112.9, 149.6, 156.1 and 231.2 ppm.

The multiplicity-edited ^{1}H–^{13}C HSQC, ^{1}H–^{13}C HMBC and ^{1}H–^{15}N HMBC spectra are given on the following pages. Use these spectra to assign each ^{1}H, ^{13}C and ^{15}N resonance to its corresponding nucleus.

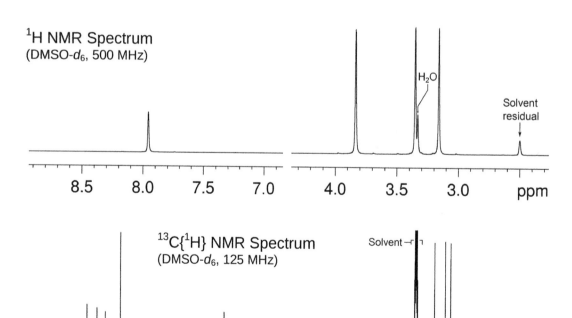

^{1}H NMR Spectrum
(DMSO-d_6, 500 MHz)

H$_2$O

Solvent residual

^{13}C{^{1}H} NMR Spectrum
(DMSO-d_6, 125 MHz)

Solvent

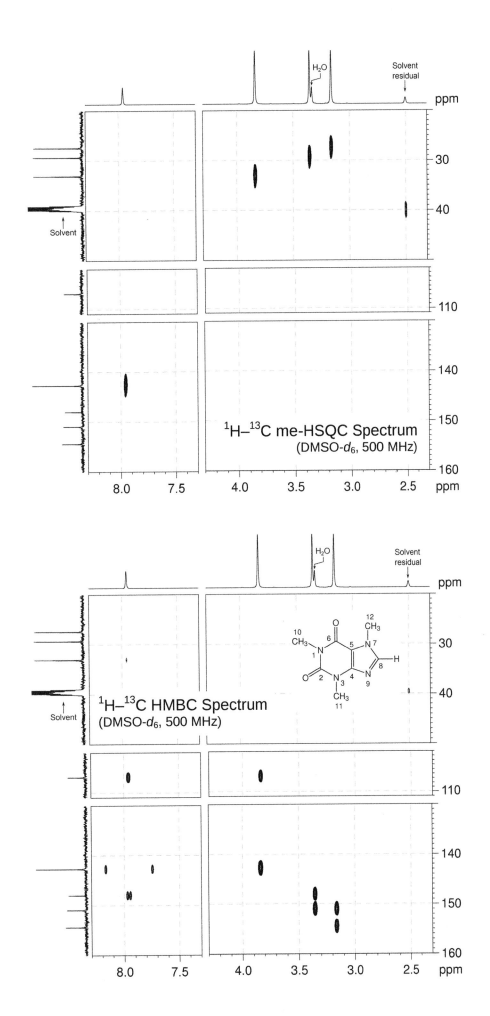

^1H–^{13}C me-HSQC Spectrum
(DMSO-d_6, 500 MHz)

^1H–^{13}C HMBC Spectrum
(DMSO-d_6, 500 MHz)

215

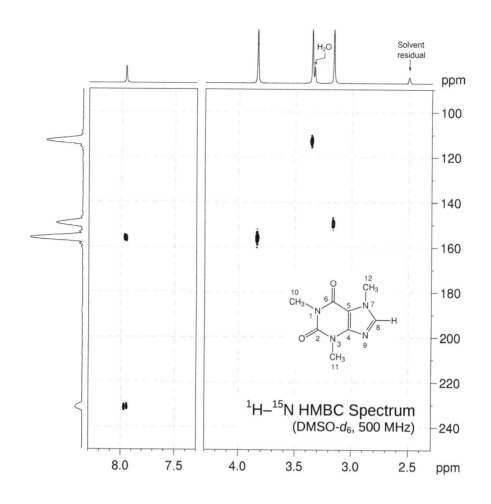

^1H–^{15}N HMBC Spectrum
(DMSO-d_6, 500 MHz)

Proton	Chemical Shift (ppm)	Nucleus	Chemical Shift (ppm)
		N_1	
		C_2	
		N_3	
		C_4	
		C_5	
		C_6	
		N_7	
H_8		C_8	
		N_9	
H_{10}		C_{10}	
H_{11}		C_{11}	
H_{12}		C_{12}	

Problem 48

Identify the following compound.

Molecular Formula: $C_{10}H_{11}NO$

IR: 2252 cm^{-1}

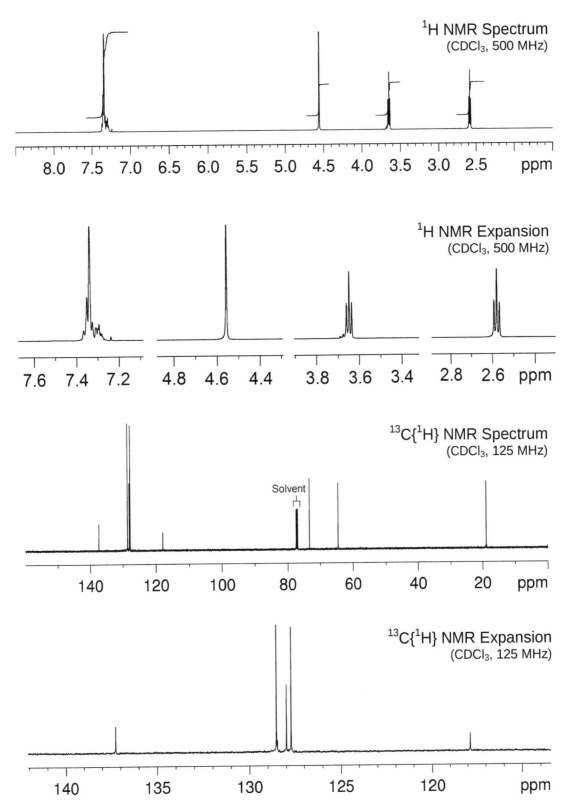

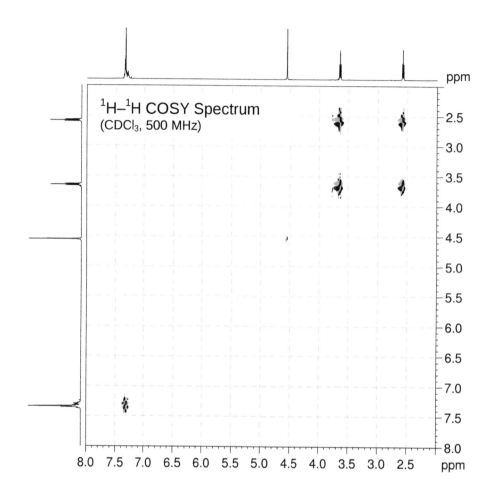

¹H–¹H COSY Spectrum
(CDCl₃, 500 MHz)

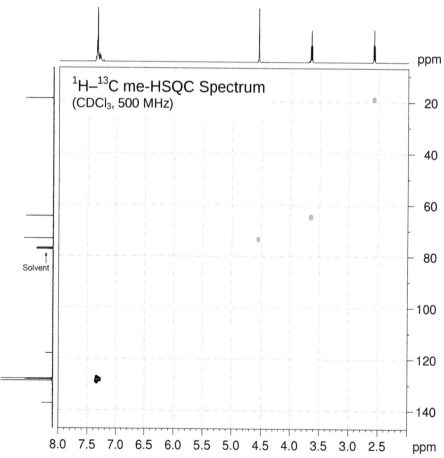

¹H–¹³C me-HSQC Spectrum
(CDCl₃, 500 MHz)

Solvent

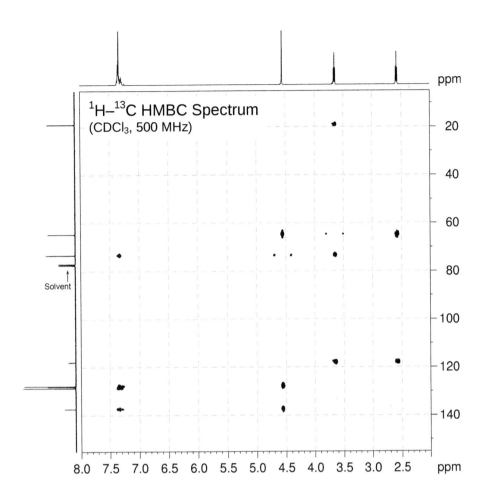

^{1}H–^{13}C HMBC Spectrum
(CDCl$_3$, 500 MHz)

Solvent

219

Problem 49

Identify the following compound.
Molecular Formula: $C_{10}H_{18}O$

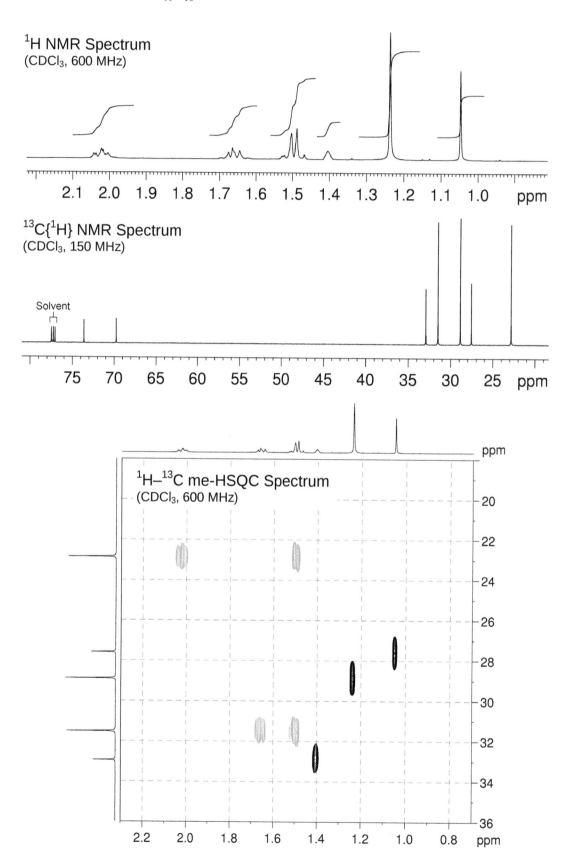

^{1}H NMR Spectrum
(CDCl$_3$, 600 MHz)

^{13}C{^{1}H} NMR Spectrum
(CDCl$_3$, 150 MHz)

Solvent

^{1}H–^{13}C me-HSQC Spectrum
(CDCl$_3$, 600 MHz)

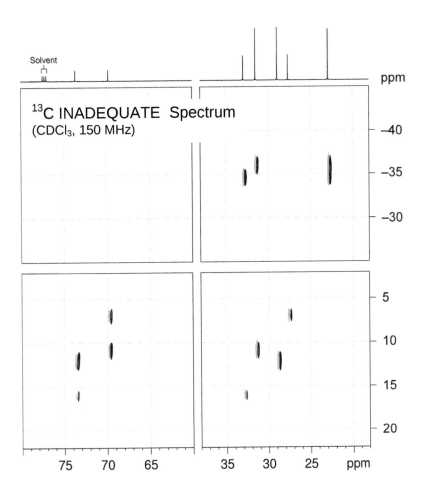

^{13}C INADEQUATE Spectrum
(CDCl$_3$, 150 MHz)

Solvent

221

Problem 50

Given below are nine benzoquinones which are isomers of $C_{10}H_{12}O_2$. The ^1H and $^{13}C\{^1H\}$ NMR spectra of one of the isomers are given below. The $^1H-^{13}C$ me-HSQC and $^1H-^{13}C$ HMBC spectra are given on the following page. To which of these compounds do the spectra belong?

HINT: In benzoquinones, three-bond C–H correlations ($^3J_{CH}$) in the HMBC spectrum are significantly larger than two-bond C–H correlations ($^2J_{CH}$) (which are negligible).

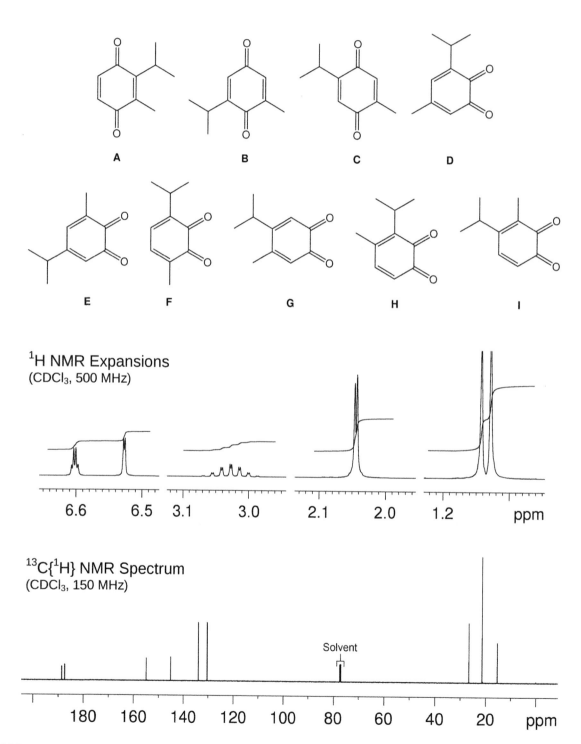

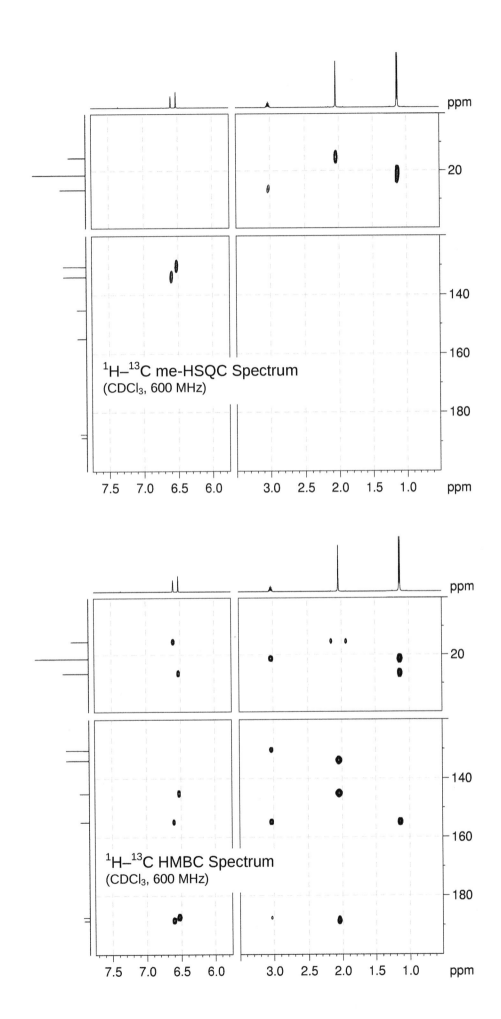

¹H–¹³C me-HSQC Spectrum
(CDCl₃, 600 MHz)

¹H–¹³C HMBC Spectrum
(CDCl₃, 600 MHz)

Problem 51

Identify the following compound.
Molecular Formula: C_9H_9BrO
IR: 3164 (br), 2953 cm^{-1}

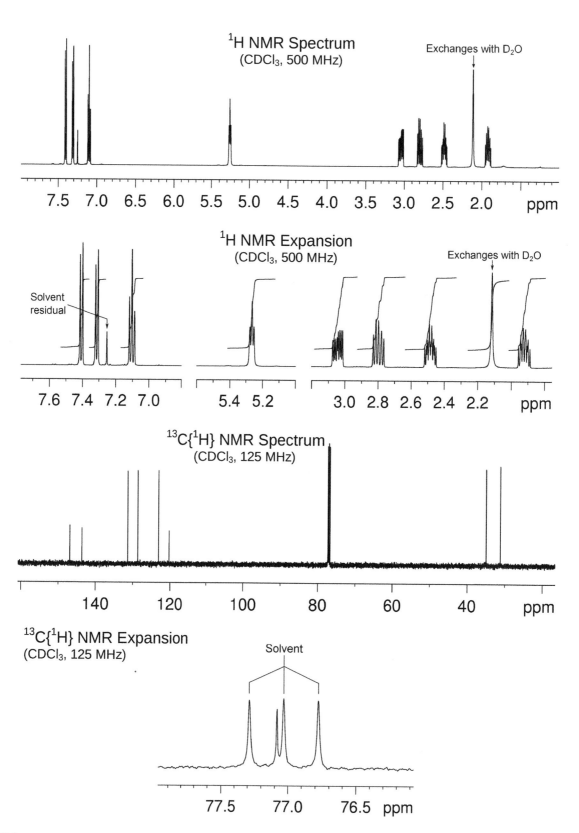

¹H NMR Spectrum
(CDCl₃, 500 MHz)

Exchanges with D₂O

¹H NMR Expansion
(CDCl₃, 500 MHz)

Exchanges with D₂O

Solvent residual

¹³C{¹H} NMR Spectrum
(CDCl₃, 125 MHz)

¹³C{¹H} NMR Expansion
(CDCl₃, 125 MHz)

Solvent

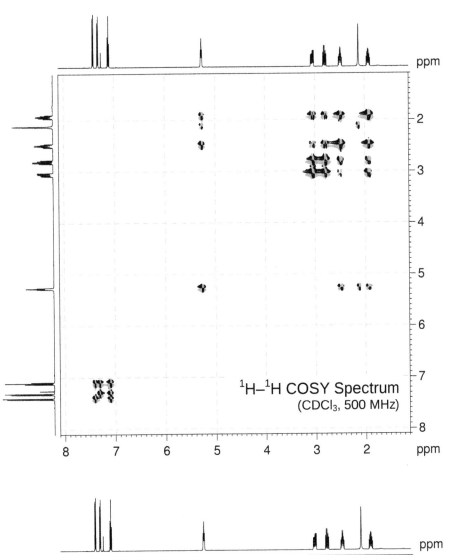

$^1H–^1H$ COSY Spectrum
(CDCl$_3$, 500 MHz)

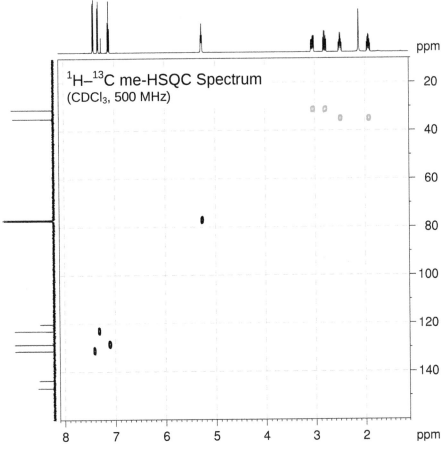

$^1H–^{13}C$ me-HSQC Spectrum
(CDCl$_3$, 500 MHz)

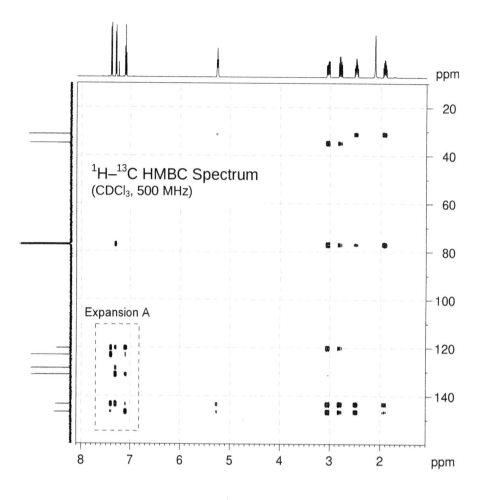

^{1}H–^{13}C HMBC Spectrum
(CDCl$_3$, 500 MHz)

Expansion A

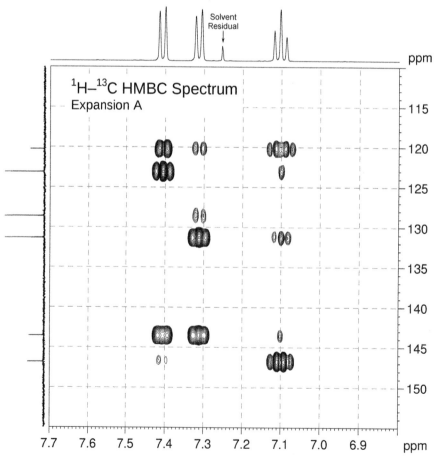

^{1}H–^{13}C HMBC Spectrum
Expansion A

Solvent
Residual

The ^1H and $^{13}C\{^1H\}$ NMR spectra of 1-bromo-4-methylnaphthalene ($C_{11}H_9Br$) recorded in CDCl$_3$ solution at 298 K and 500 MHz are given below.

The ^1H NMR spectrum has signals at δ 2.58, 7.08, 7.50, 7.54, 7.61, 7.91 and 8.23 ppm.

The $^{13}C\{^1H\}$ NMR spectrum has signals at δ 19.2, 120.6, 124.5, 126.4, 126.85, 126.91, 127.6, 129.4, 131.7, 133.7 and 134.3 ppm.

The 2D ^1H–^1H COSY, multiplicity-edited ^1H–^{13}C HSQC and ^1H–^{13}C HMBC spectra are given on the following pages. Use these spectra to assign the ^1H and $^{13}C\{^1H\}$ resonances for this compound.

Proton	Chemical Shift (ppm)	Carbon	Chemical Shift (ppm)
		C_1	
H_2		C_2	
H_3		C_3	
		C_4	
		C_5	
H_6		C_6	
H_7		C_7	
H_8		C_8	
H_9		C_9	
		C_{10}	
H_{11}		C_{11}	

^{1}H NMR Spectrum
(CDCl$_3$, 500 MHz)

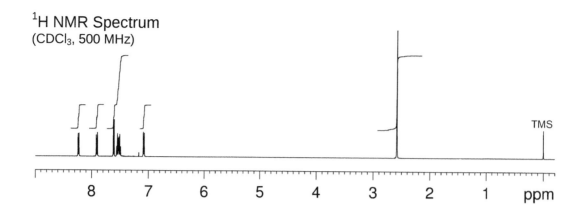

^{1}H NMR Expansion
(CDCl$_3$, 500 MHz)

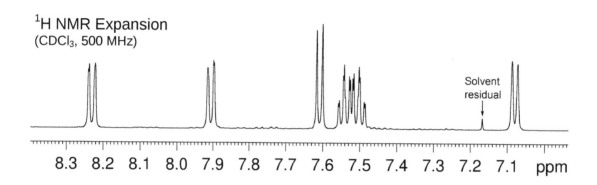

^{13}C{^{1}H} NMR Spectrum
(CDCl$_3$, 125 MHz)

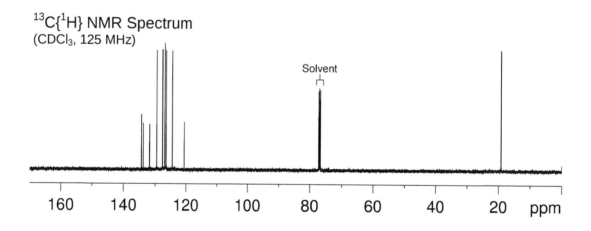

^{13}C{^{1}H} NMR Expansion
(CDCl$_3$, 125 MHz)

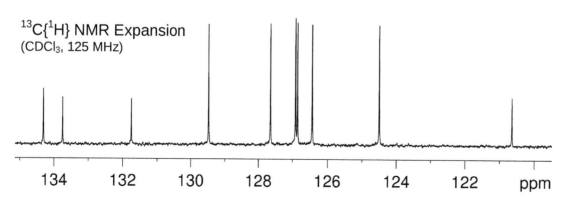

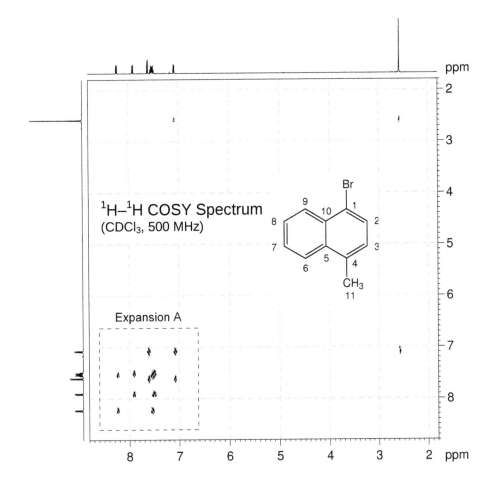

^{1}H–^{1}H COSY Spectrum
(CDCl$_3$, 500 MHz)

Expansion A

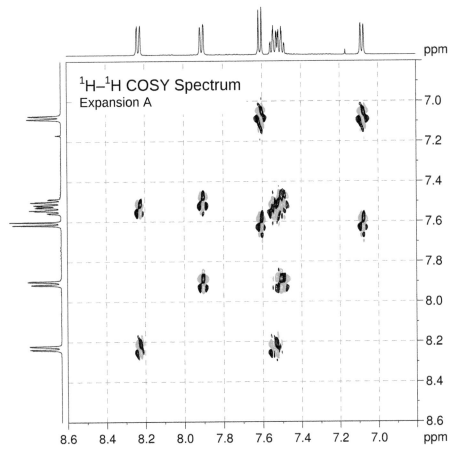

^{1}H–^{1}H COSY Spectrum
Expansion A

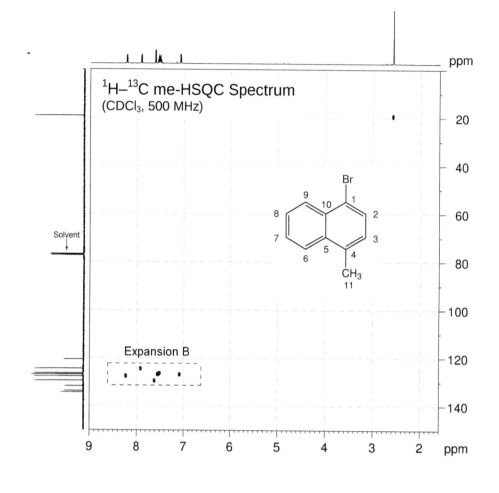

¹H–¹³C me-HSQC Spectrum
(CDCl₃, 500 MHz)

Solvent

Expansion B

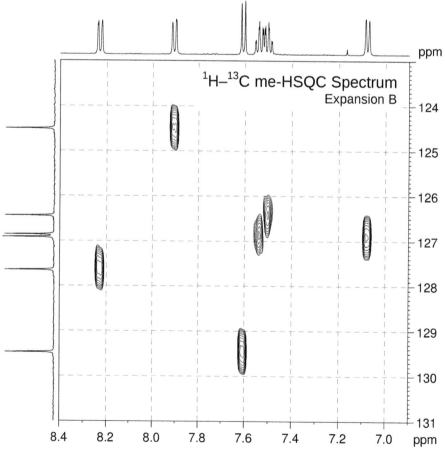

¹H–¹³C me-HSQC Spectrum
Expansion B

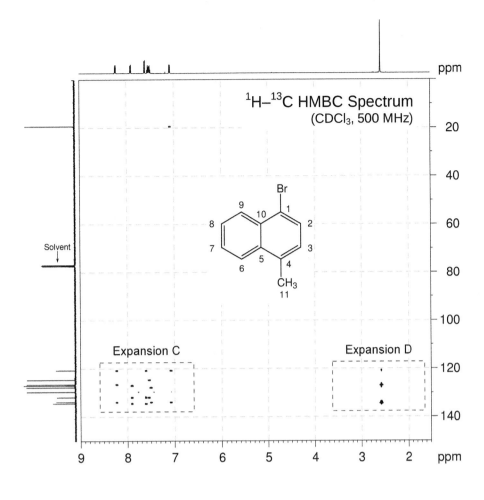

^1H–^{13}C HMBC Spectrum
(CDCl$_3$, 500 MHz)

Solvent

Expansion C

Expansion D

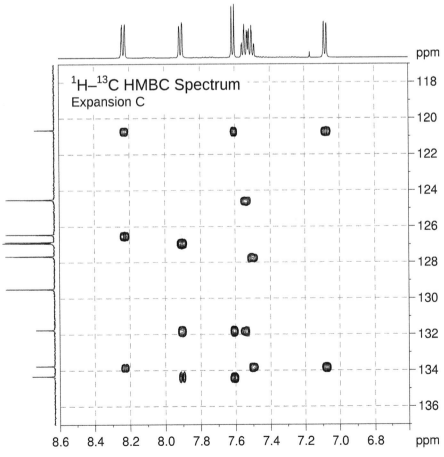

^1H–^{13}C HMBC Spectrum
Expansion C

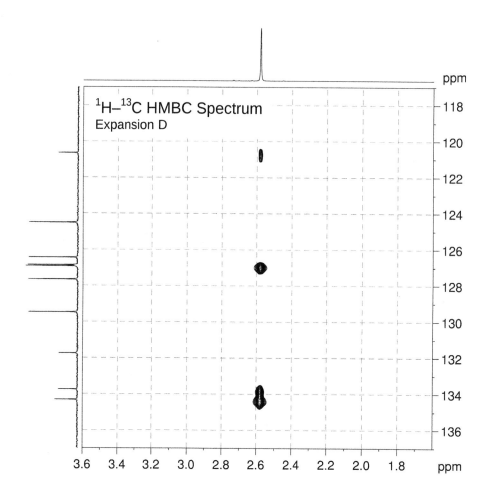

Problem 53

Identify the following compound.
Molecular Formula: $C_{10}H_{14}O$
IR: 3377 (br) cm^{-1}

^{1}H NMR Spectrum
(CDCl$_3$, 400 MHz)

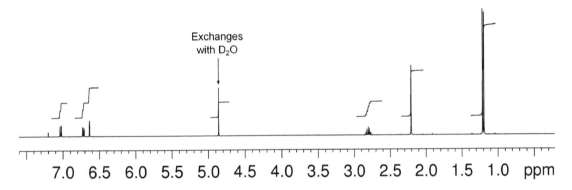

^{1}H NMR Expansion Spectrum
(CDCl$_3$, 400 MHz)

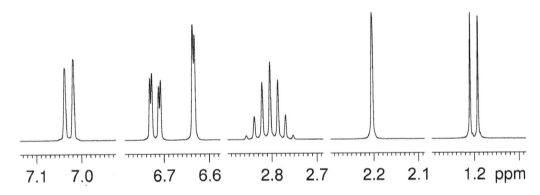

^{13}C{^{1}H} NMR Spectrum
(CDCl$_3$, 100 MHz)

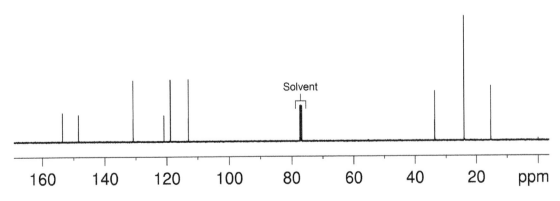

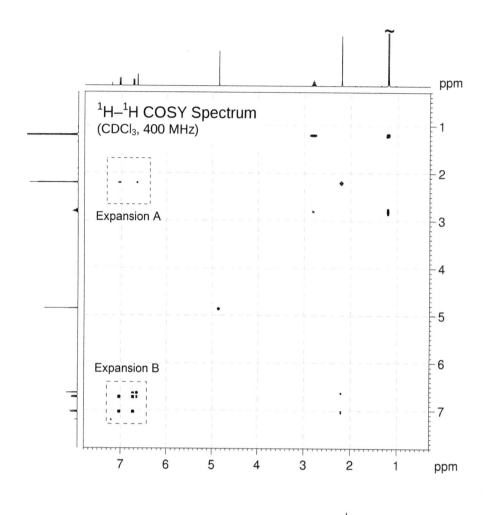

¹H–¹H COSY Spectrum
(CDCl₃, 400 MHz)

Expansion A

Expansion B

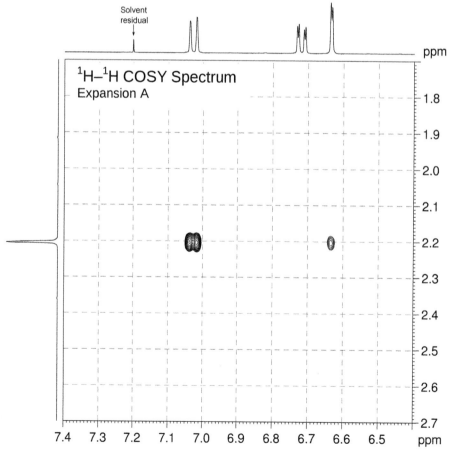

¹H–¹H COSY Spectrum
Expansion A

Solvent
residual

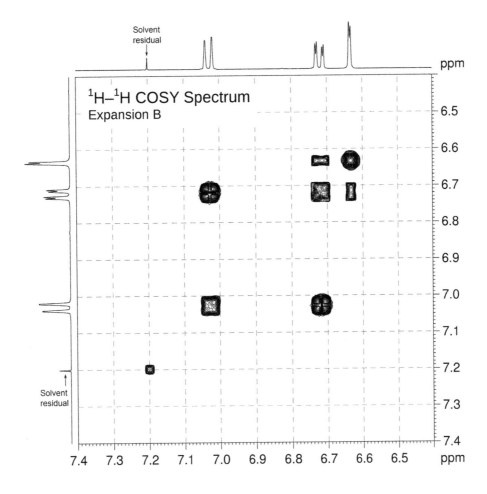

^{1}H–^{1}H COSY Spectrum
Expansion B

Solvent residual

Solvent residual

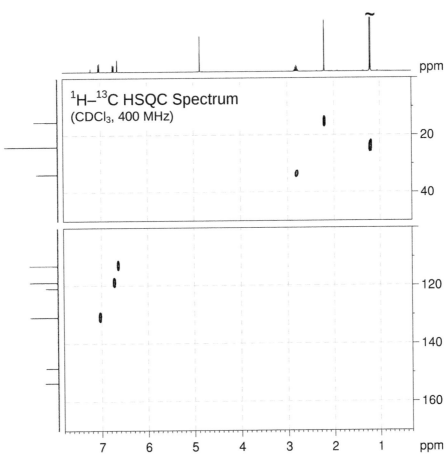

^{1}H–^{13}C HSQC Spectrum
(CDCl$_3$, 400 MHz)

235

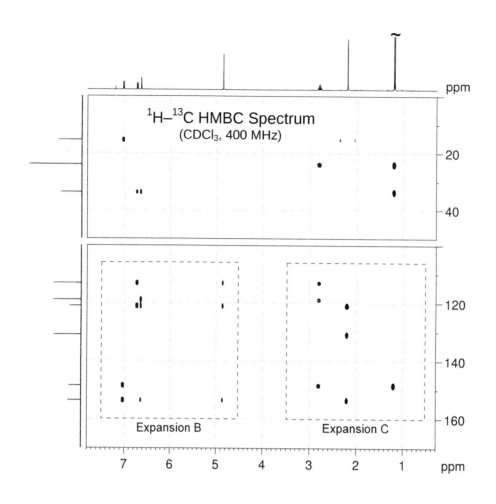

^{1}H–^{13}C HMBC Spectrum
(CDCl$_3$, 400 MHz)

Expansion B   Expansion C

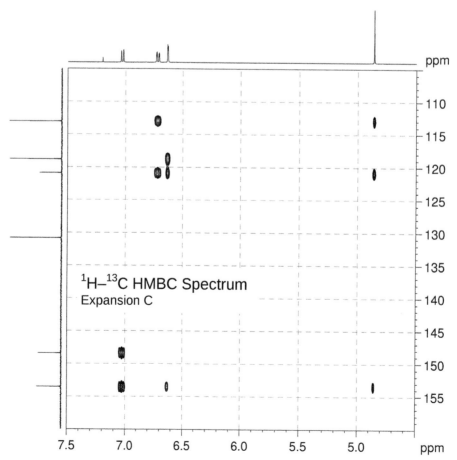

^{1}H–^{13}C HMBC Spectrum
Expansion C

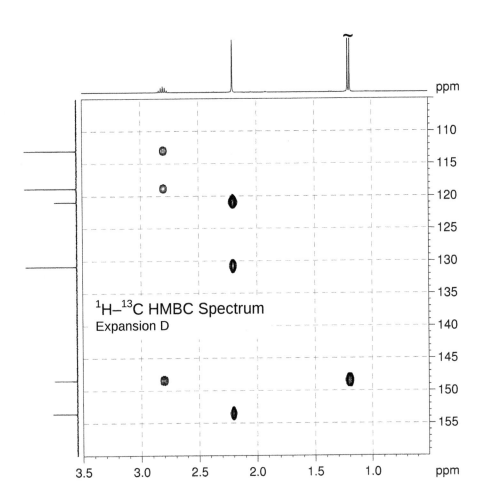

$^1H-^{13}C$ HMBC Spectrum
Expansion D

237

Problem 54

This compound is an amide, and amides frequently exist as a mixture of stereoisomers due to restricted rotation about the C–N bond.

Use the spectra below to identify the compound, including the stereochemistry of the major isomer.

Molecular Formula: $C_{10}H_{11}NO_2$

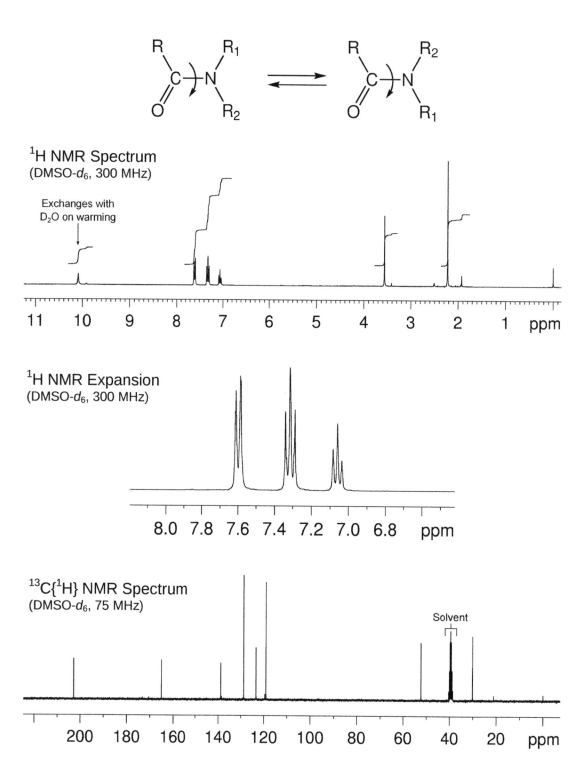

^{1}H NMR Spectrum
(DMSO-d_6, 300 MHz)

Exchanges with
D$_2$O on warming

11   10   9   8   7   6   5   4   3   2   1   ppm

^{1}H NMR Expansion
(DMSO-d_6, 300 MHz)

8.0 7.8 7.6 7.4 7.2 7.0 6.8   ppm

^{13}C{^{1}H} NMR Spectrum
(DMSO-d_6, 75 MHz)

Solvent

200   180   160   140   120   100   80   60   40   20   ppm

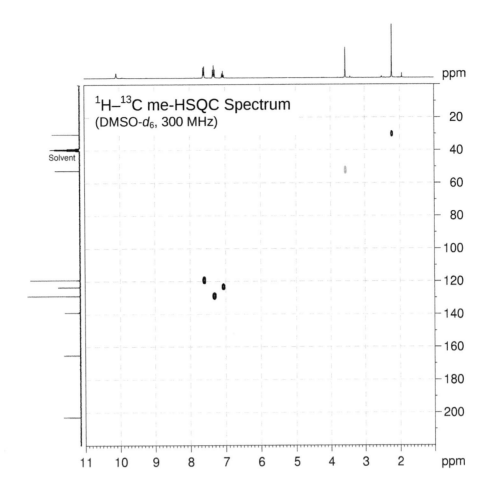

¹H–¹³C me-HSQC Spectrum
(DMSO-d_6, 300 MHz)

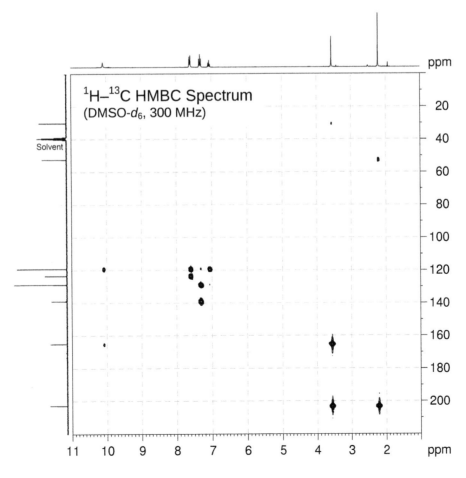

¹H–¹³C HMBC Spectrum
(DMSO-d_6, 300 MHz)

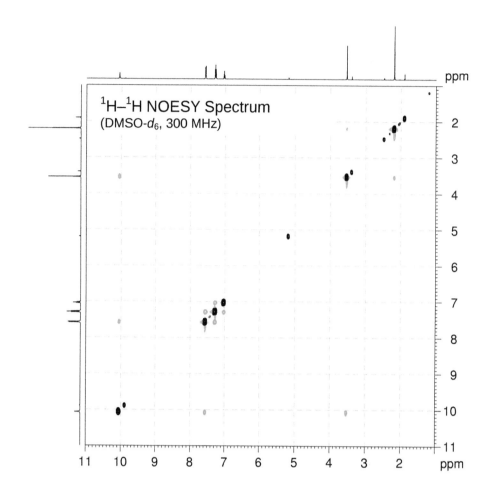

^{1}H–^{1}H NOESY Spectrum
(DMSO-d_6, 300 MHz)

Problem 55

Identify the following compound.
Molecular Formula: $C_7H_{10}N_2O_3$
IR: 2930, 2258, 1740, 1698, 1643 cm^{-1}

^{1}H NMR Spectrum
(CDCl$_3$/DMSO-d_6, 500 MHz)

Exchanges with
D$_2$O on warming

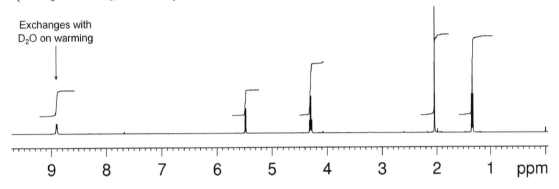

^{1}H NMR Expansion Spectrum
(CDCl$_3$/DMSO-d_6, 500 MHz)

Exchanges with
D$_2$O on warming

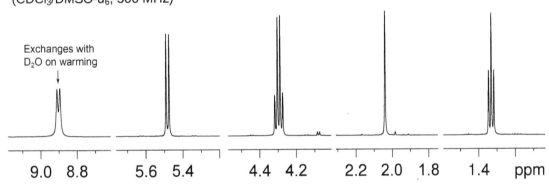

^{13}C{^{1}H} NMR Spectrum
(CDCl$_3$/DMSO-d_6, 125 MHz)

Solvent

Solvent

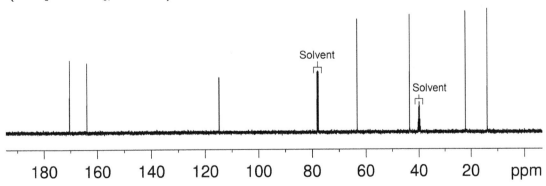

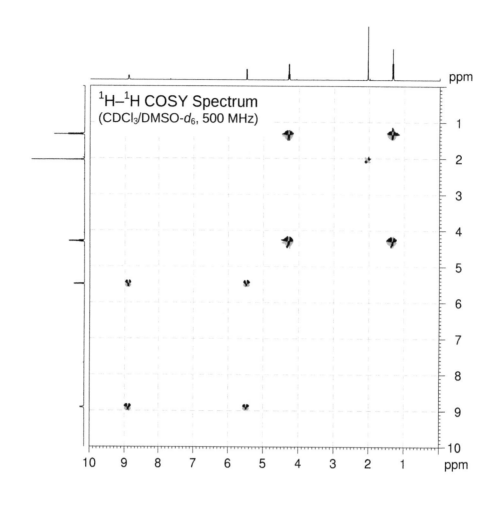

¹H–¹H COSY Spectrum
(CDCl₃/DMSO-d_6, 500 MHz)

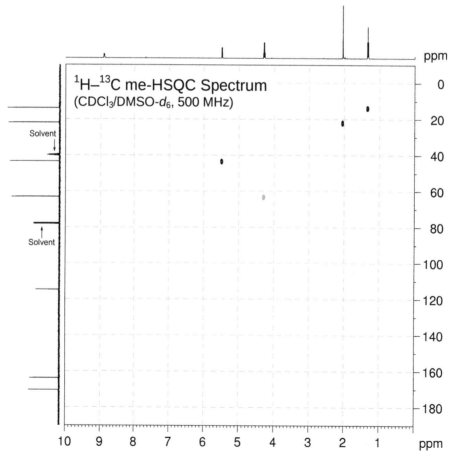

¹H–¹³C me-HSQC Spectrum
(CDCl₃/DMSO-d_6, 500 MHz)

Solvent

Solvent

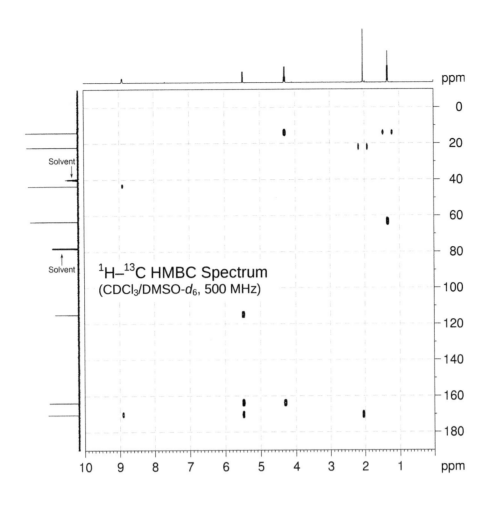

^1H–^{13}C HMBC Spectrum
(CDCl$_3$/DMSO-d_6, 500 MHz)

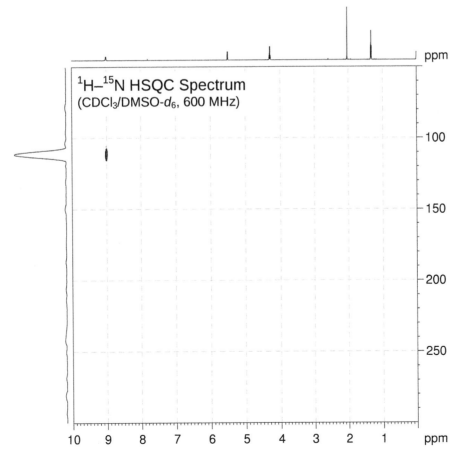

^1H–^{15}N HSQC Spectrum
(CDCl$_3$/DMSO-d_6, 600 MHz)

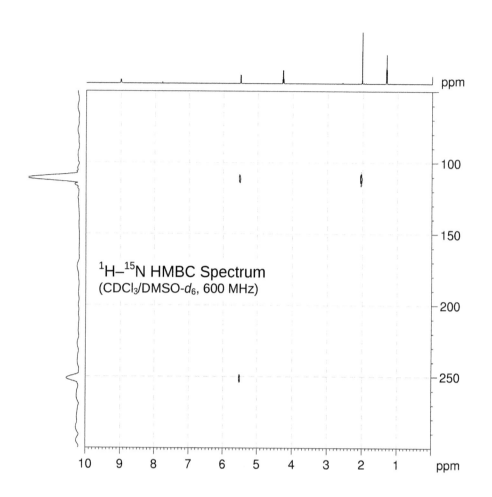

^{1}H–^{15}N HMBC Spectrum
(CDCl$_3$/DMSO-d_6, 600 MHz)

Problem 56

The following spectra belong to a macrocyclic sesquiterpene which contains three *trans*-substituted double bonds. Identify the compound.

Molecular Formula: $C_{15}H_{24}$

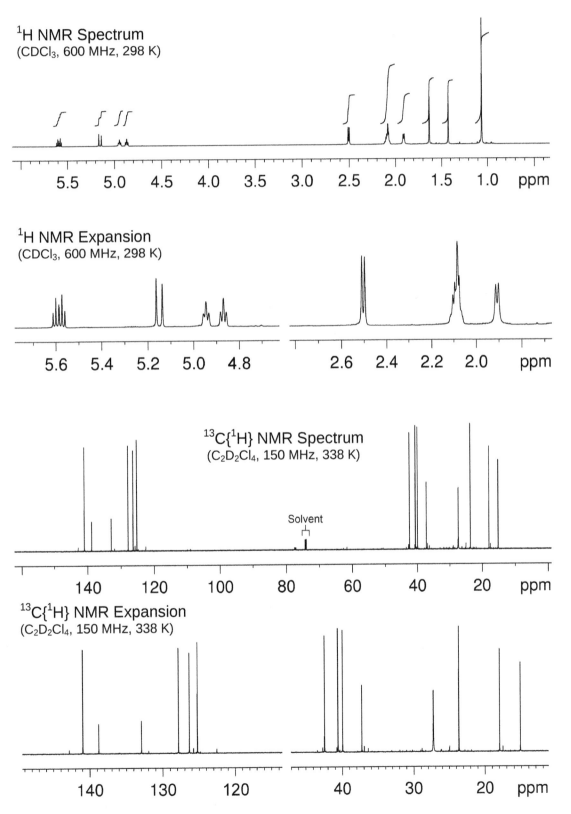

¹H NMR Spectrum
(CDCl₃, 600 MHz, 298 K)

¹H NMR Expansion
(CDCl₃, 600 MHz, 298 K)

¹³C{¹H} NMR Spectrum
(C₂D₂Cl₄, 150 MHz, 338 K)

Solvent

¹³C{¹H} NMR Expansion
(C₂D₂Cl₄, 150 MHz, 338 K)

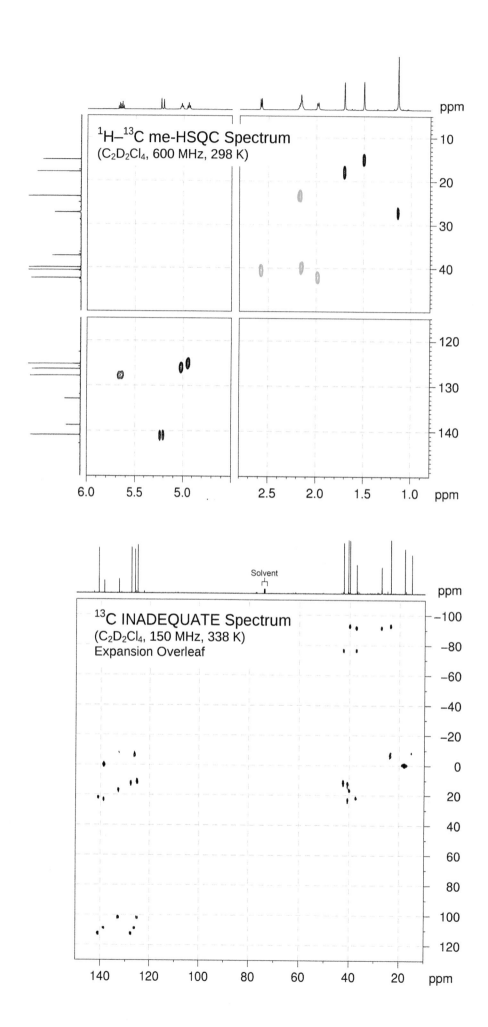

^{1}H–^{13}C me-HSQC Spectrum
(C$_2$D$_2$Cl$_4$, 600 MHz, 298 K)

^{13}C INADEQUATE Spectrum
(C$_2$D$_2$Cl$_4$, 150 MHz, 338 K)
Expansion Overleaf

Solvent

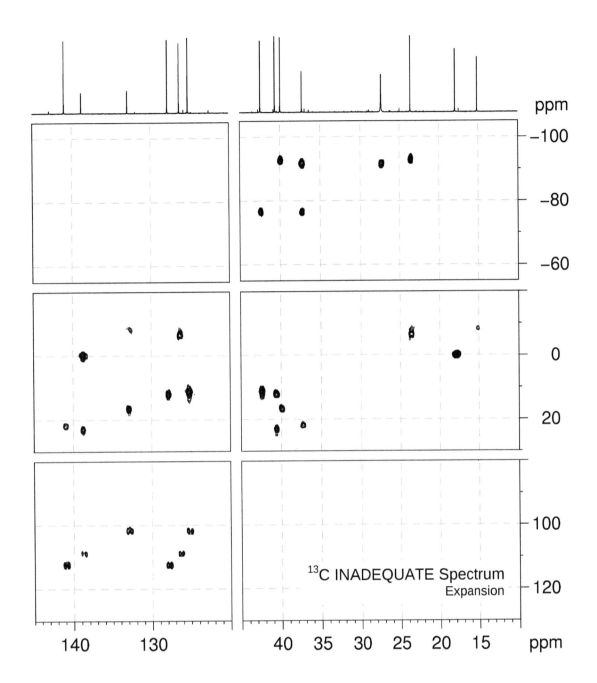

^{13}C INADEQUATE Spectrum
Expansion

247

Problem 57

Identify the following compound.
Molecular Formula: $C_{10}H_{10}O_3$
IR: 3300–2400 (br), 1685 cm^{-1}

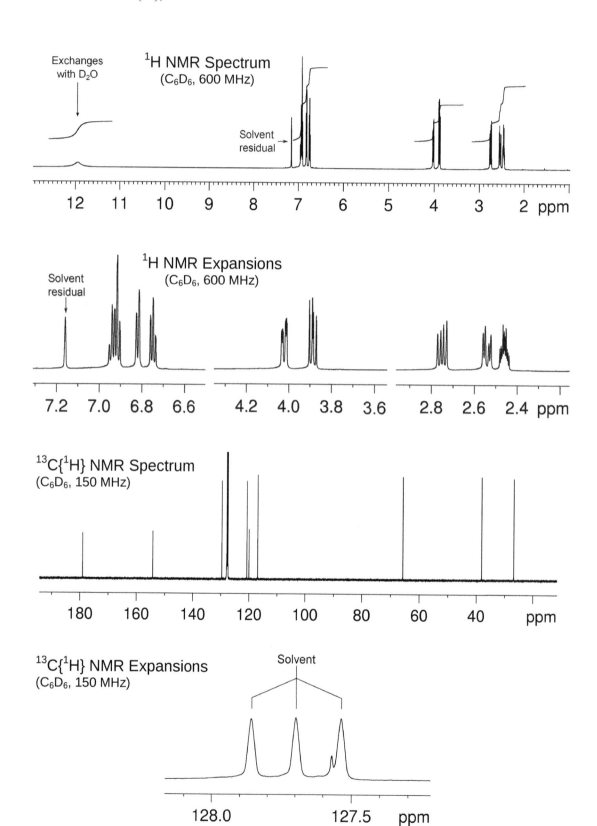

^{1}H NMR Spectrum
(C$_6$D$_6$, 600 MHz)

Exchanges with D$_2$O

Solvent residual

^{1}H NMR Expansions
(C$_6$D$_6$, 600 MHz)

Solvent residual

^{13}C{^{1}H} NMR Spectrum
(C$_6$D$_6$, 150 MHz)

^{13}C{^{1}H} NMR Expansions
(C$_6$D$_6$, 150 MHz)

Solvent

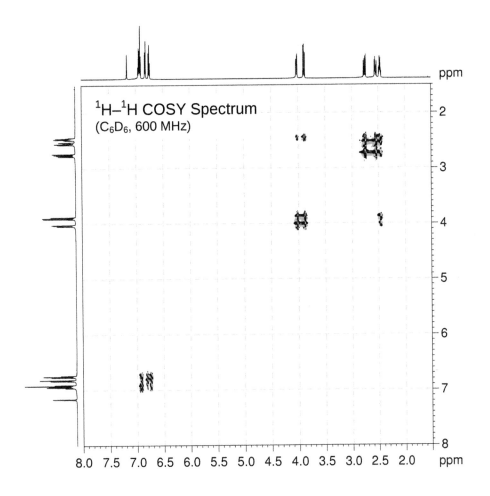

^1H–^1H COSY Spectrum
(C$_6$D$_6$, 600 MHz)

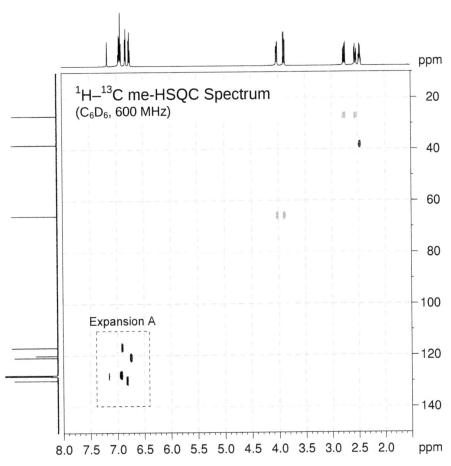

^1H–^{13}C me-HSQC Spectrum
(C$_6$D$_6$, 600 MHz)

Expansion A

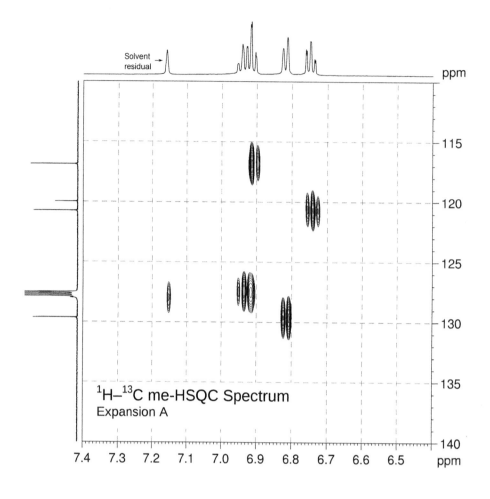

^{1}H–^{13}C me-HSQC Spectrum
Expansion A

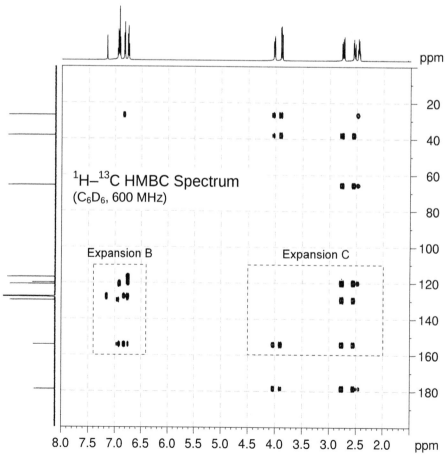

^{1}H–^{13}C HMBC Spectrum
(C$_6$D$_6$, 600 MHz)

Expansion B

Expansion C

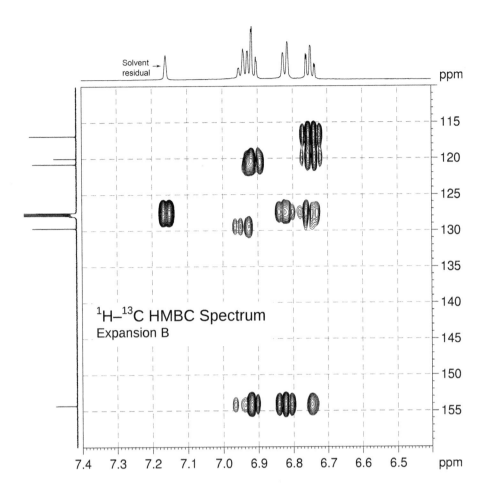

^{1}H–^{13}C HMBC Spectrum
Expansion B

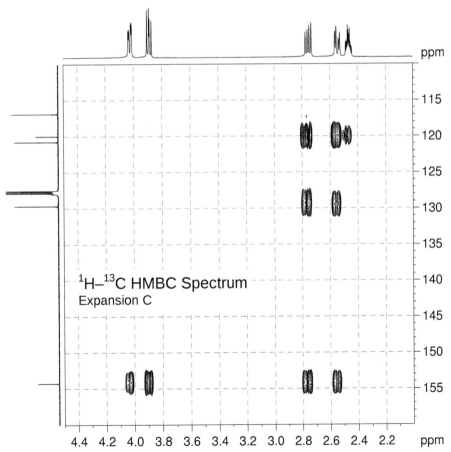

^{1}H–^{13}C HMBC Spectrum
Expansion C

Problem 58

The ^1H and $^{13}C\{^1H\}$ NMR spectra of quinidine ($C_{20}H_{24}N_2O_2$) recorded in DMSO-d_6 solution at 298 K and 600 MHz are given below.

The ^1H NMR spectrum has signals at δ 1.38 (m, 1H), 1.43 (m, 2H), 1.68 (m, 1H), 1.94 (m, 1H), 2.17 (m, 1H), 2.53 (m, 1H), 2.63 (m, 1H), 2.70 (m, 1H), 3.00 (m, 1H), 3.06 (m, 1H), 3.90 (s, 3H), 5.05 (m, 1H), 5.08 (m, 1H), 5.32 (dd, J = 4.7, 6.2 Hz, 1H), 5.73 (d, J = 4.7 Hz, 1H), 6.10 (ddd, J = 7.5, 10.2, 17.5 Hz, 1H), 7.39 (dd, J = 2.8, 9.3 Hz, 1H), 7.48 (d, J = 2.8 Hz, 1H), 7.53 (d, J = 4.4 Hz, 1H), 7.94 (d, J = 9.3 Hz, 1H) and 8.69 (d, J = 4.4 Hz, 1H) ppm.

The $^{13}C\{^1H\}$ NMR spectrum has signals at δ 23.1, 26.2, 27.8, 39.8, 48.4, 49.1, 55.3, 60.5, 70.7, 102.3, 114.2, 118.9, 120.9, 126.9, 131.0, 141.2, 143.8, 147.4, 149.4 and 156.7 ppm.

The 2D ^1H–^1H COSY, multiplicity-edited ^1H–^{13}C HSQC, ^1H–^{13}C HMBC and ^1H–^1H NOESY spectra are given on the following pages. Use these spectra to assign the ^1H and $^{13}C\{^1H\}$ resonances for this compound.

¹H NMR Spectrum
(DMSO-d_6, 600 MHz)

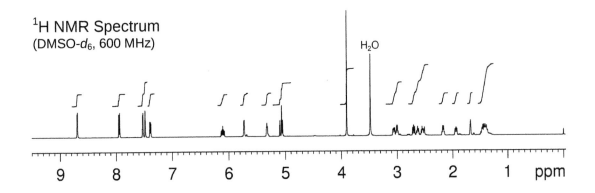

H₂O

¹H NMR Expansion Spectrum
(DMSO-d_6, 600 MHz)

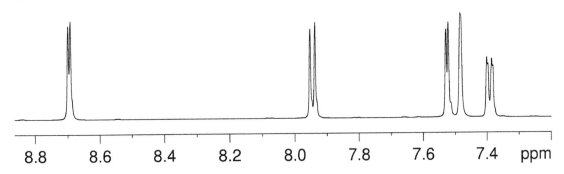

¹H NMR Expansion Spectrum
(DMSO-d_6, 600 MHz)

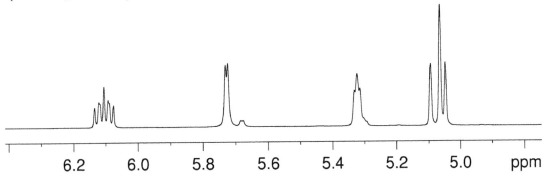

¹H NMR Expansion Spectrum
(DMSO-d_6, 600 MHz)

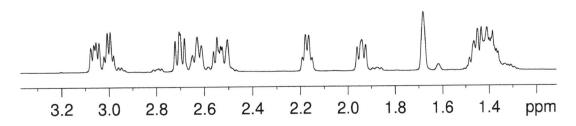

^{13}C{^{1}H} NMR Spectrum
(DMSO-d_6, 150 MHz)

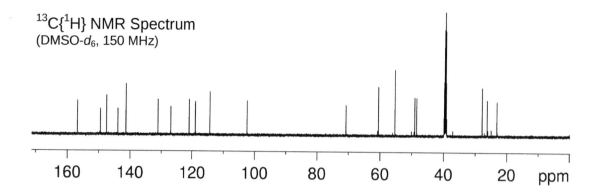

^{13}C{^{1}H} NMR Expansion Spectrum
(DMSO-d_6, 150 MHz)

Solvent

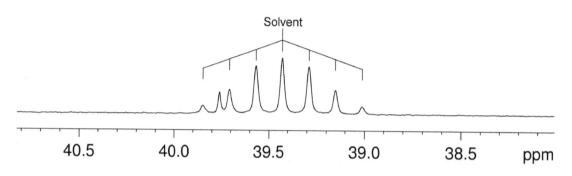

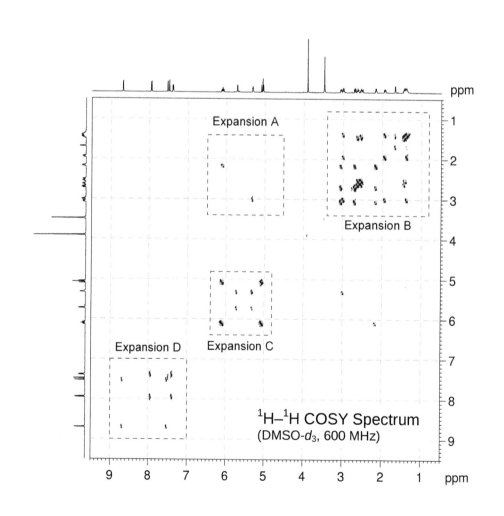

Expansion A

Expansion B

Expansion D   Expansion C

^{1}H–^{1}H COSY Spectrum
(DMSO-d_3, 600 MHz)

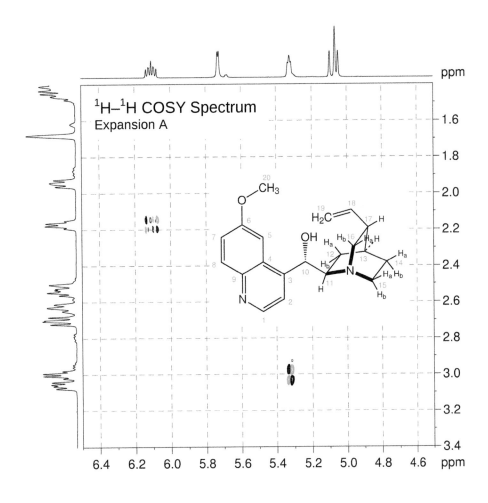

^{1}H–^{1}H COSY Spectrum
Expansion A

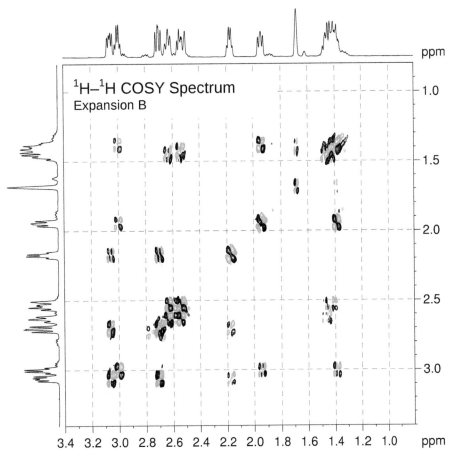

^{1}H–^{1}H COSY Spectrum
Expansion B

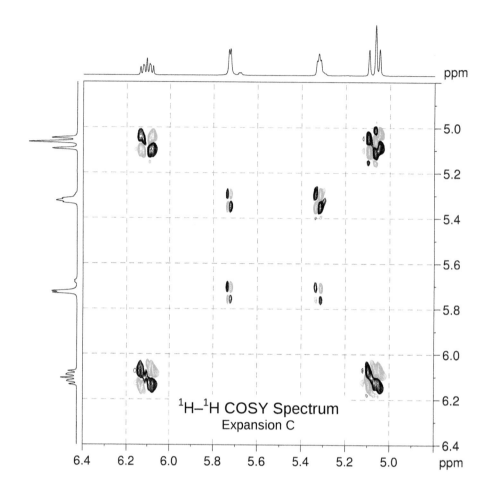

^1H–^1H COSY Spectrum
Expansion C

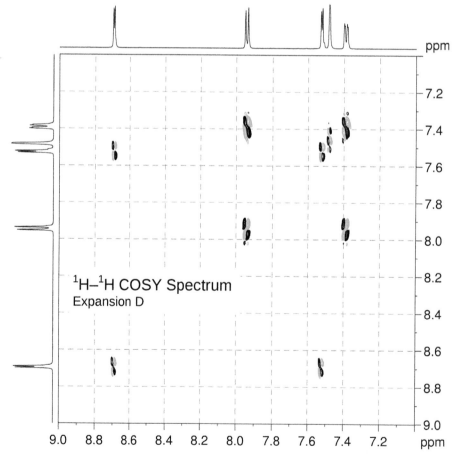

^1H–^1H COSY Spectrum
Expansion D

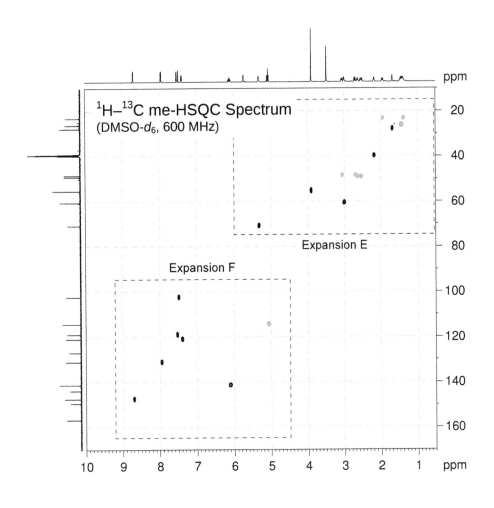

¹H–¹³C me-HSQC Spectrum
(DMSO-d₆, 600 MHz)

Expansion E

Expansion F

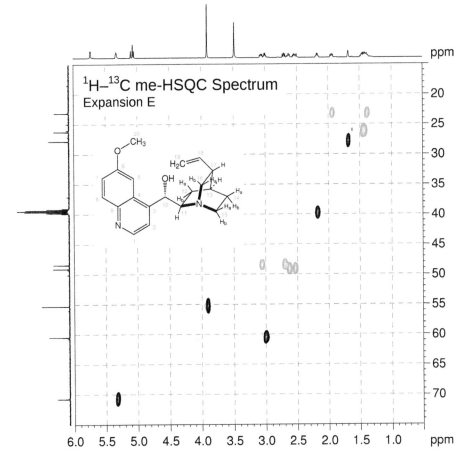

¹H–¹³C me-HSQC Spectrum
Expansion E

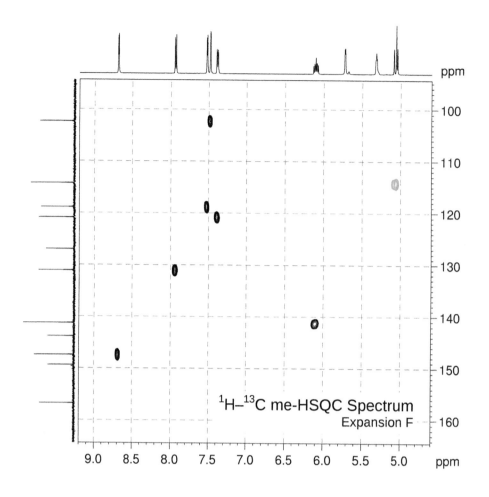

^{1}H–^{13}C me-HSQC Spectrum
Expansion F

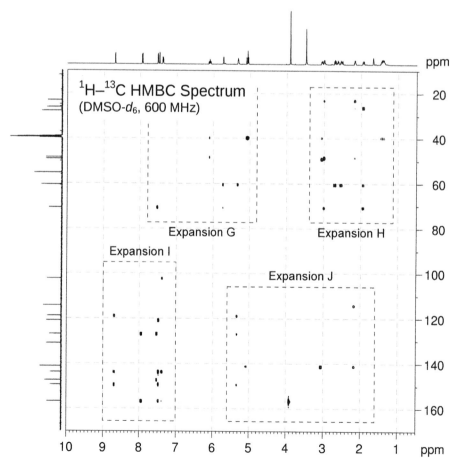

^{1}H–^{13}C HMBC Spectrum
(DMSO-d_6, 600 MHz)

Expansion G

Expansion H

Expansion I

Expansion J

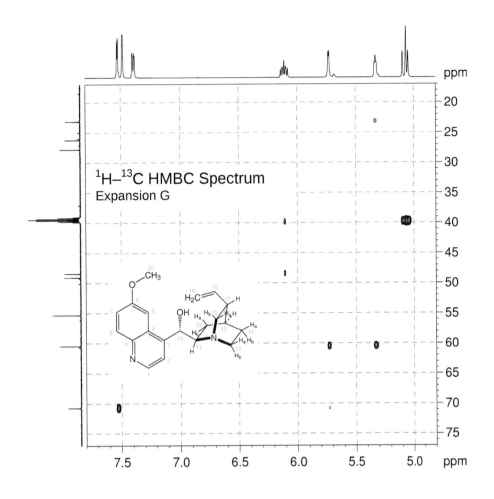

¹H–¹³C HMBC Spectrum
Expansion G

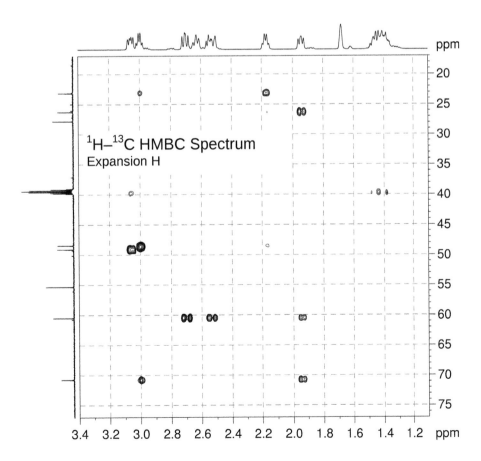

¹H–¹³C HMBC Spectrum
Expansion H

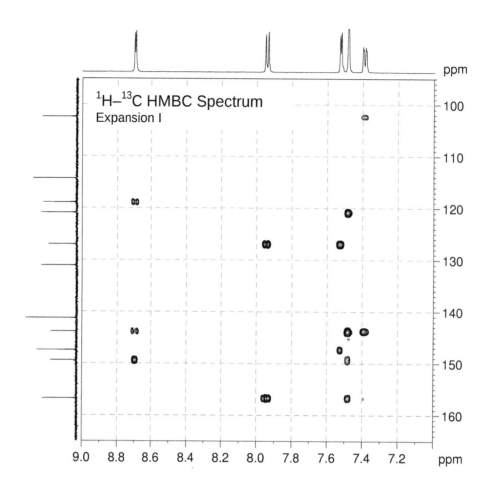

¹H–¹³C HMBC Spectrum
Expansion I

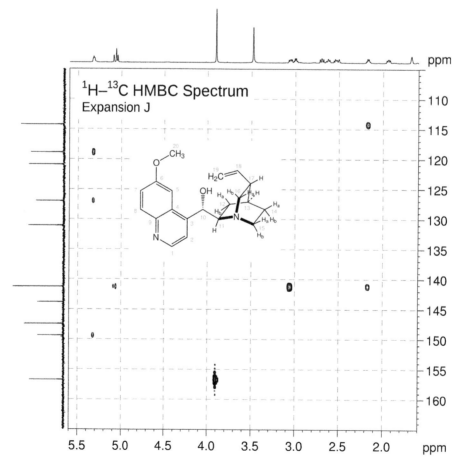

¹H–¹³C HMBC Spectrum
Expansion J

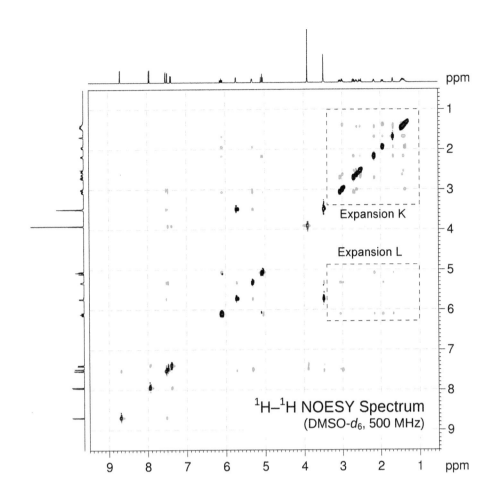

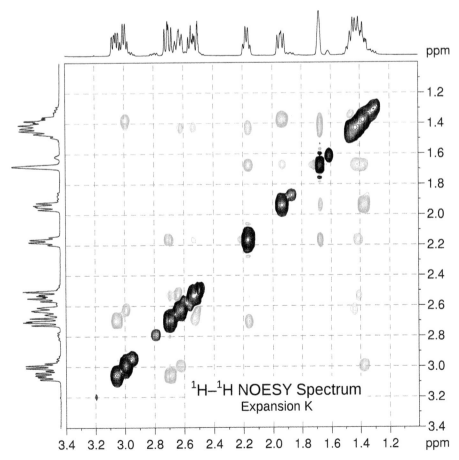

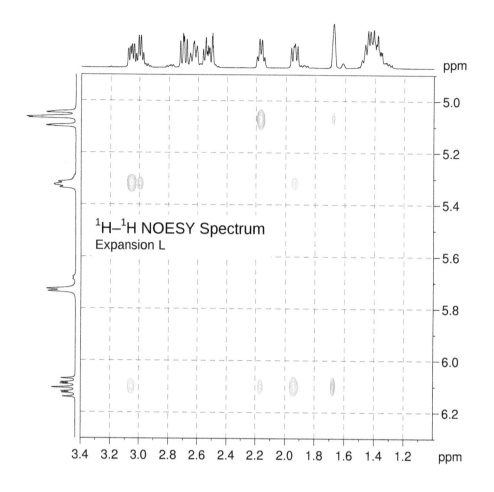

^{1}H–^{1}H NOESY Spectrum
Expansion L

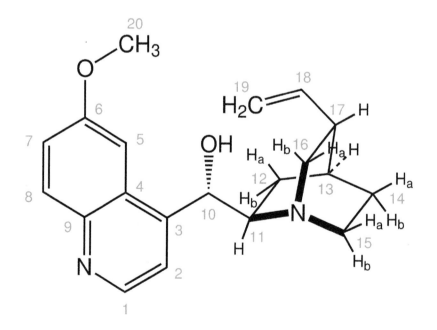

Proton	Chemical Shift (ppm)	Carbon	Chemical Shift (ppm)
H_1		C_1	
H_2		C_2	
		C_3	
		C_4	
H_5		C_5	
		C_6	
H_7		C_7	
H_8		C_8	
		C_9	
H_{10}		C_{10}	
H_{11}		C_{11}	
H_{12a}		C_{12}	
H_{12b}			
H_{13}		C_{13}	
H_{14a}		C_{14}	
H_{14b}			
H_{15a}		C_{15}	
H_{15b}			
H_{16a}		C_{16}	
H_{16b}			
H_{17}		C_{17}	
H_{18}		C_{18}	
H_{19a}		C_{19}	
H_{19b}			
H_{20}		C_{20}	
OH			

Problem 59

Identify the following compound.

Molecular Formula: $C_{13}H_{21}NO_3$

HINT: This compound contains a secondary amine.

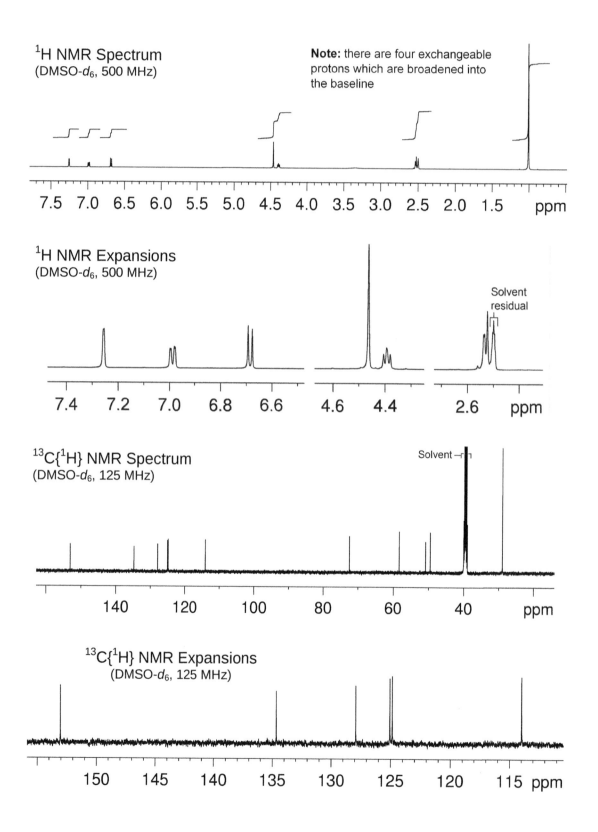

^{1}H NMR Spectrum
(DMSO-d_6, 500 MHz)

Note: there are four exchangeable protons which are broadened into the baseline

^{1}H NMR Expansions
(DMSO-d_6, 500 MHz)

Solvent residual

^{13}C{^{1}H} NMR Spectrum
(DMSO-d_6, 125 MHz)

Solvent

^{13}C{^{1}H} NMR Expansions
(DMSO-d_6, 125 MHz)

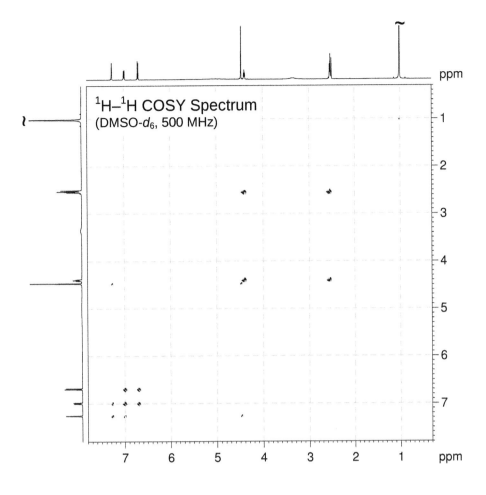

¹H–¹H COSY Spectrum
(DMSO-*d*₆, 500 MHz)

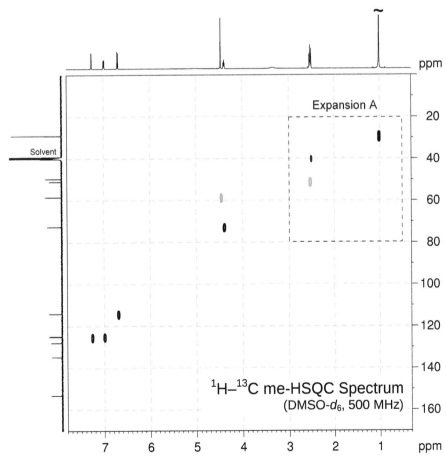

Expansion A

Solvent

¹H–¹³C me-HSQC Spectrum
(DMSO-*d*₆, 500 MHz)

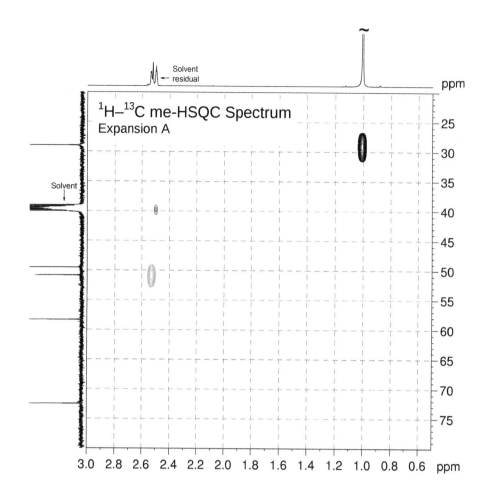

^1H–^{13}C me-HSQC Spectrum
Expansion A

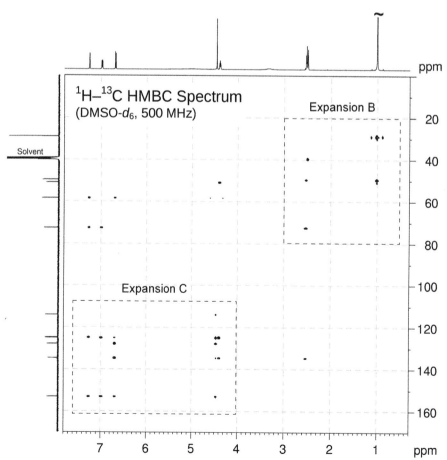

^1H–^{13}C HMBC Spectrum
(DMSO-d_6, 500 MHz)

Expansion B

Solvent

Expansion C

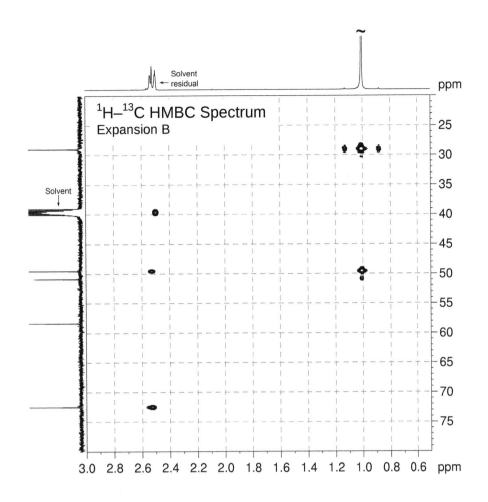

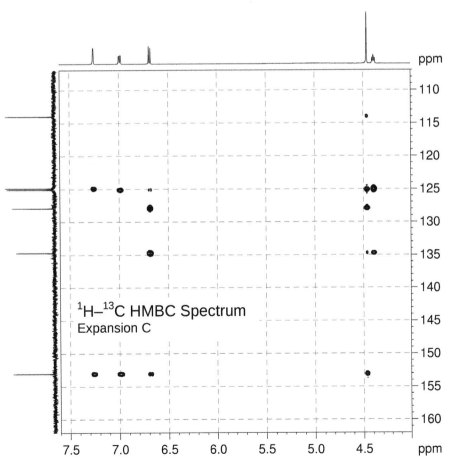

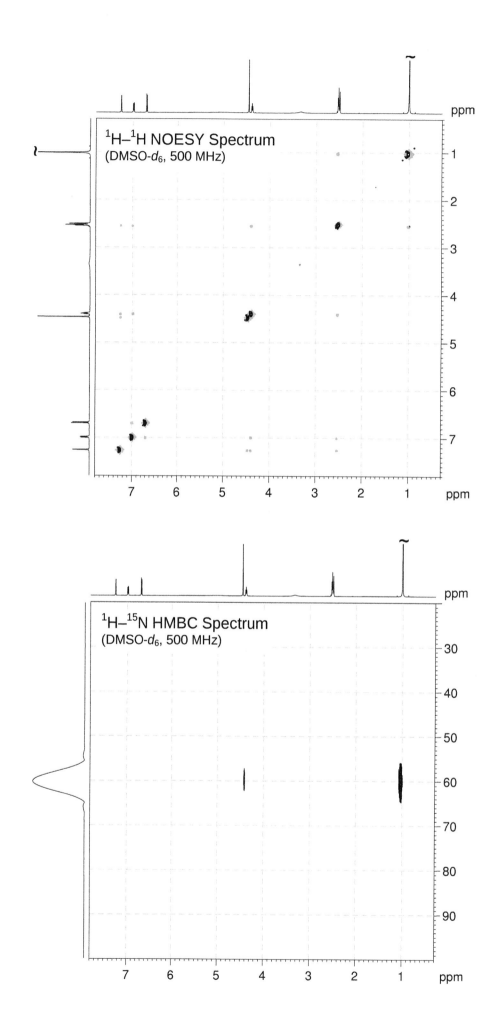

$^1H–^1H$ NOESY Spectrum
(DMSO-d_6, 500 MHz)

$^1H–^{15}N$ HMBC Spectrum
(DMSO-d_6, 500 MHz)

Problem 60

Identify the following compound.

Molecular Formula: $C_{11}H_8O_2$

IR: 2957, 2925, 2855, 1645, 1631, 1621 cm^{-1}

^{1}H NMR Spectrum
(CDCl$_3$, 600 MHz)

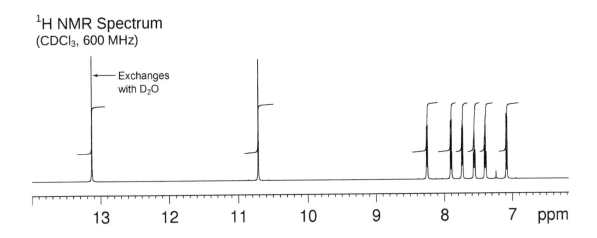

^{1}H NMR Expansion
(CDCl$_3$, 600 MHz)

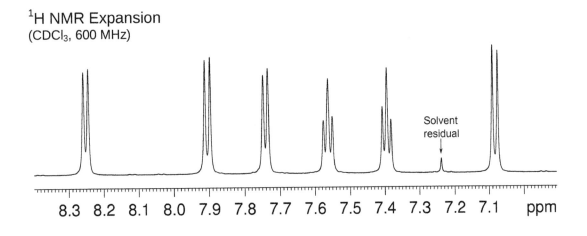

^{13}C{^{1}H} NMR Spectrum
(CDCl$_3$, 150 MHz)

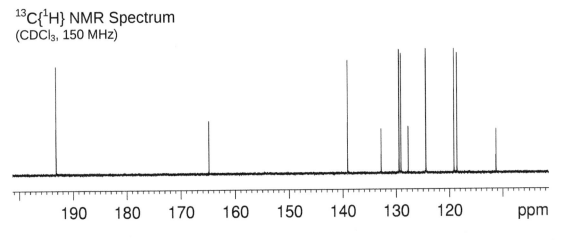

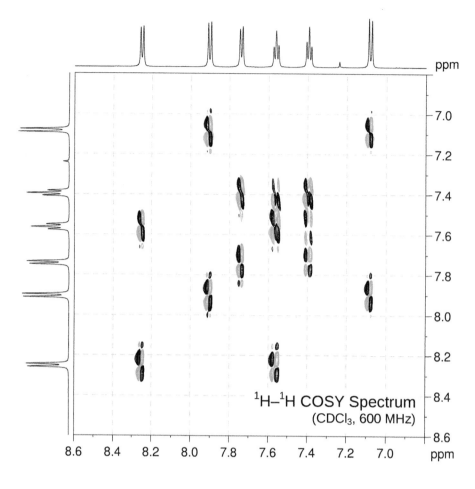

^{1}H–^{1}H COSY Spectrum
(CDCl$_3$, 600 MHz)

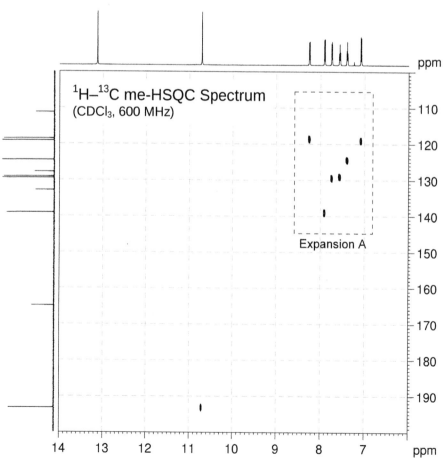

^{1}H–^{13}C me-HSQC Spectrum
(CDCl$_3$, 600 MHz)

Expansion A

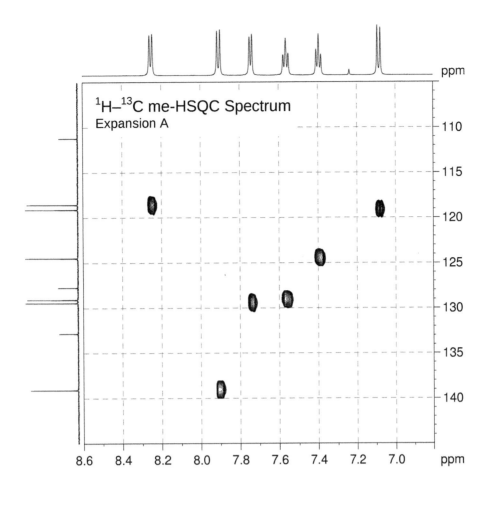

¹H–¹³C me-HSQC Spectrum
Expansion A

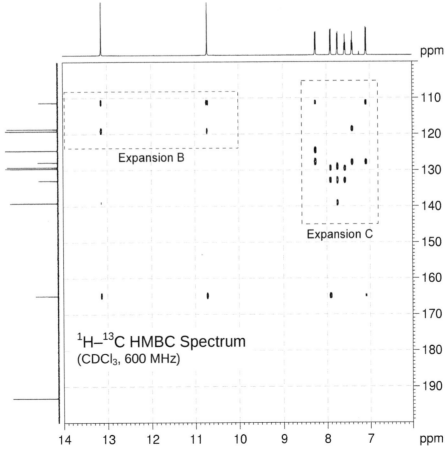

Expansion B

Expansion C

¹H–¹³C HMBC Spectrum
(CDCl₃, 600 MHz)

271

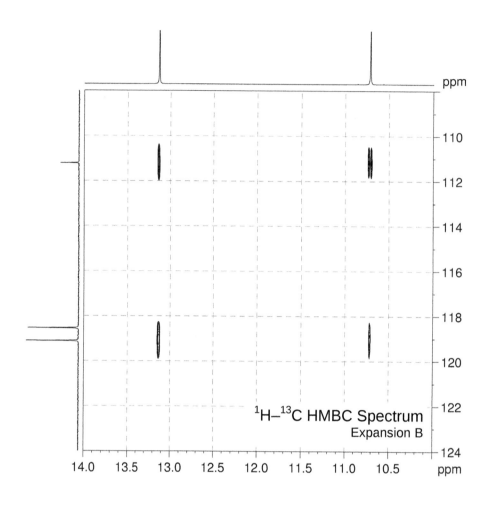

^{1}H–^{13}C HMBC Spectrum
Expansion B

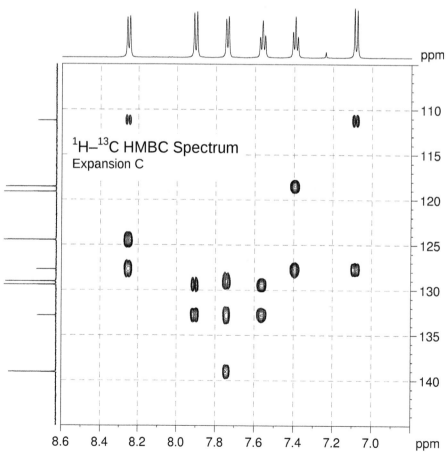

^{1}H–^{13}C HMBC Spectrum
Expansion C

Problem 61

Identify the following compound.

Molecular Formula: $C_{10}H_{10}O_2$

IR: 1689 cm^{-1}

^{1}H NMR Spectrum
(CDCl$_3$, 500 MHz)

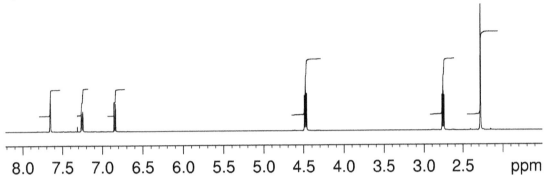

^{1}H NMR Expansions
(CDCl$_3$, 500 MHz)

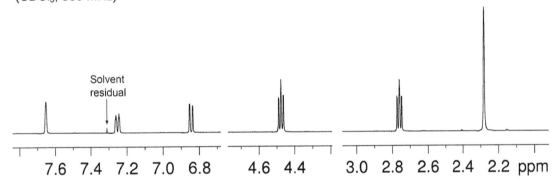

^{13}C{^{1}H} NMR Spectrum
(CDCl$_3$, 150 MHz)

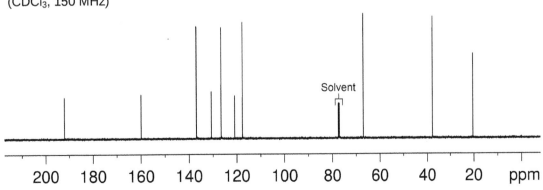

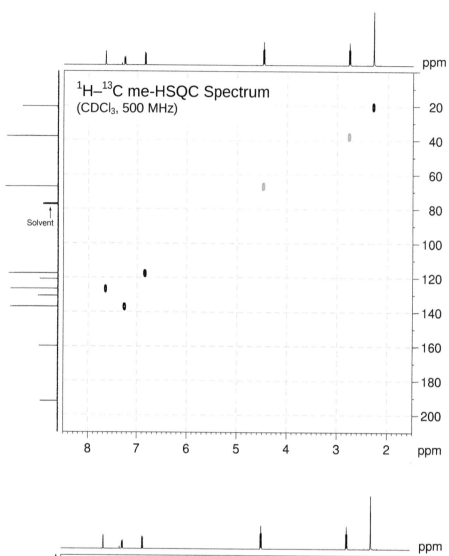

^{1}H–^{13}C me-HSQC Spectrum
(CDCl$_3$, 500 MHz)

Solvent

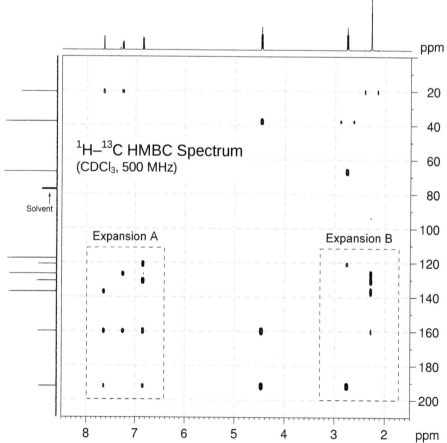

^{1}H–^{13}C HMBC Spectrum
(CDCl$_3$, 500 MHz)

Solvent

Expansion A

Expansion B

274

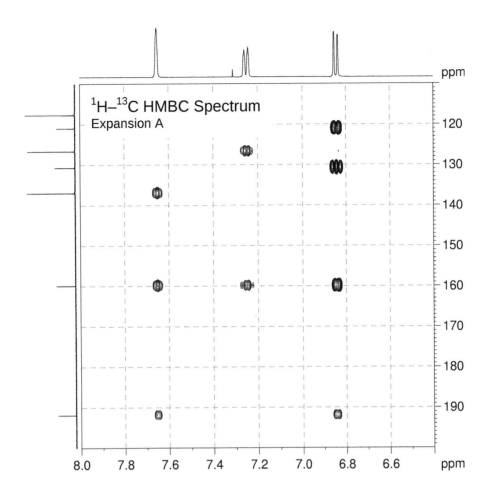

^{1}H–^{13}C HMBC Spectrum
Expansion A

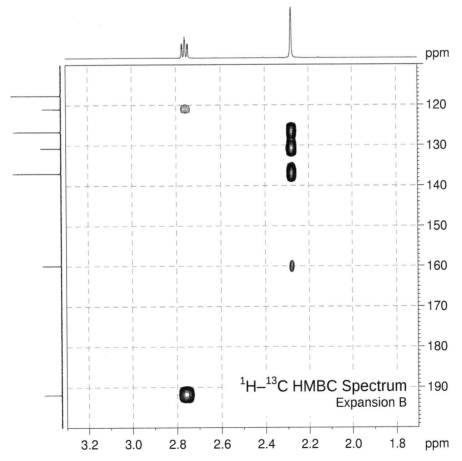

^{1}H–^{13}C HMBC Spectrum
Expansion B

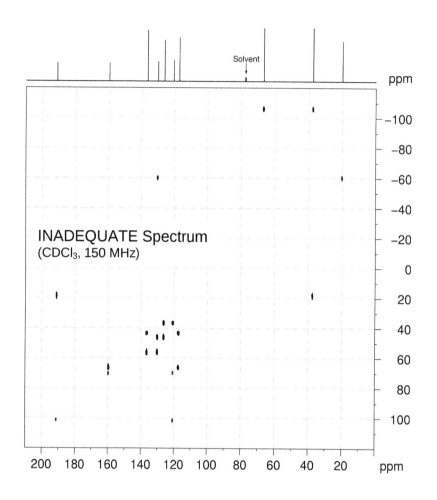

INADEQUATE Spectrum
(CDCl₃, 150 MHz)

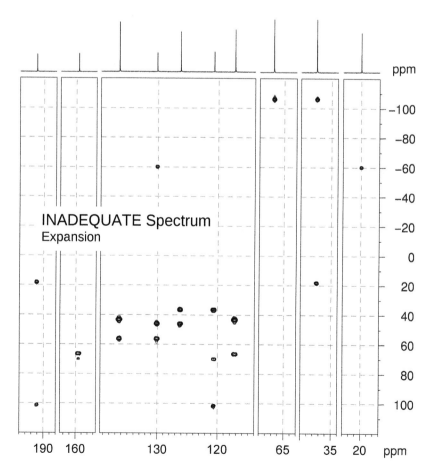

INADEQUATE Spectrum
Expansion

276

Problem 62

Identify the following compound.
Molecular Formula: $C_{10}H_{18}O$
IR: 1727 (s) cm^{-1}

^{1}H NMR Spectrum
(CDCl$_3$, 400 MHz)

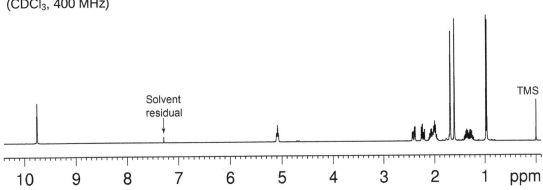

^{1}H NMR Expansion Spectrum
(CDCl$_3$, 400 MHz)

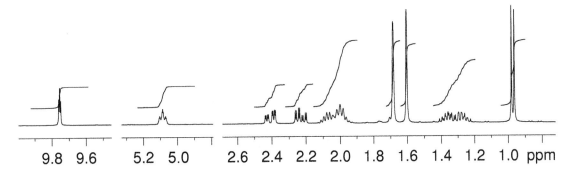

^{13}C{^{1}H} NMR Spectrum
(CDCl$_3$, 100 MHz)

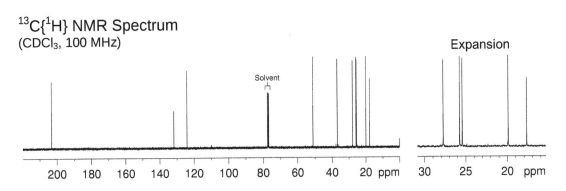

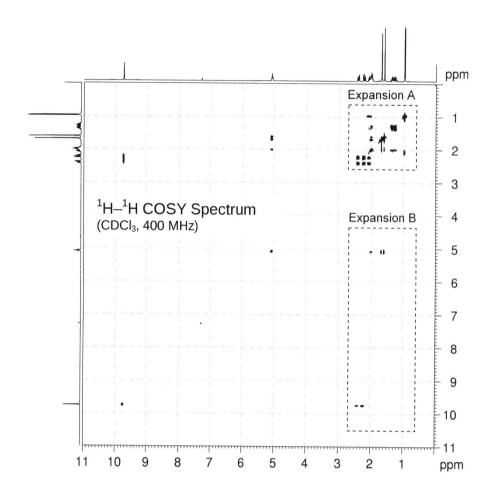

^{1}H–^{1}H COSY Spectrum
(CDCl$_3$, 400 MHz)

Expansion A

Expansion B

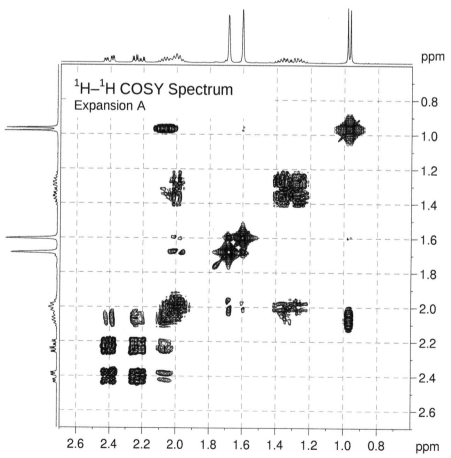

^{1}H–^{1}H COSY Spectrum
Expansion A

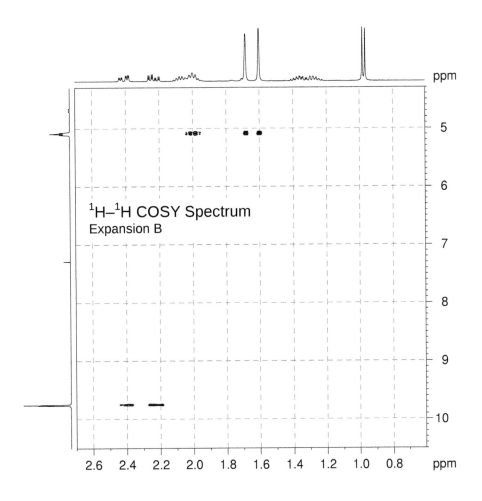

^1H–^1H COSY Spectrum

Expansion B

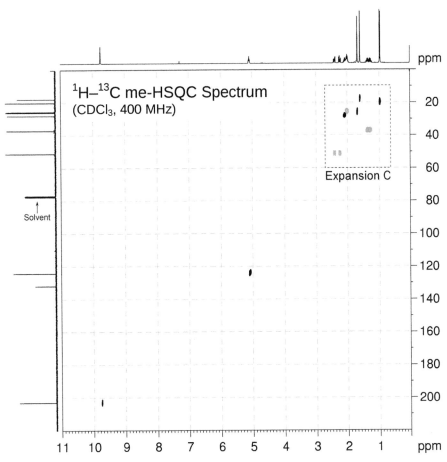

^1H–^{13}C me-HSQC Spectrum

(CDCl$_3$, 400 MHz)

Solvent

Expansion C

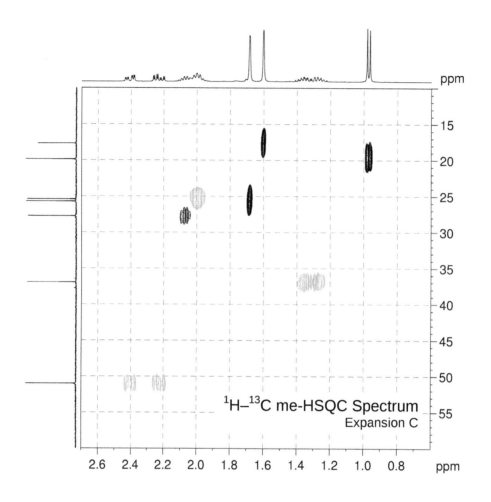

$^1H–^{13}C$ me-HSQC Spectrum
Expansion C

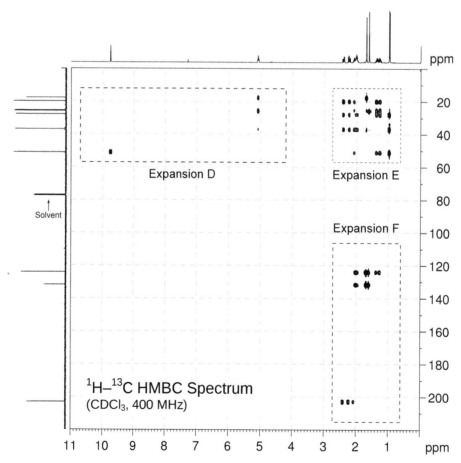

$^1H–^{13}C$ HMBC Spectrum
(CDCl$_3$, 400 MHz)

Expansion D

Expansion E

Expansion F

Solvent

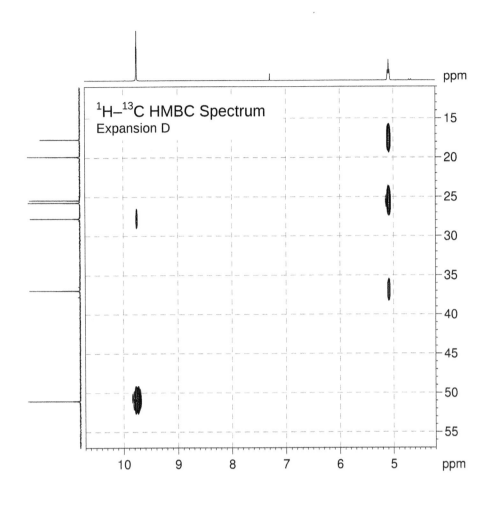

¹H–¹³C HMBC Spectrum
Expansion D

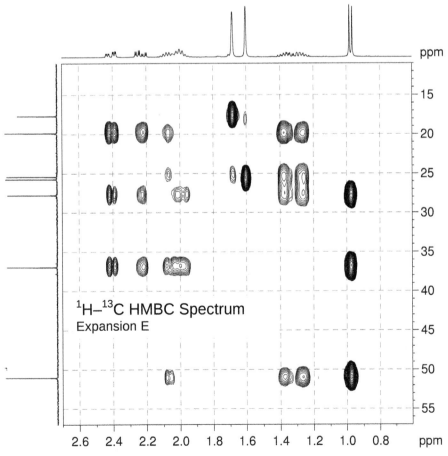

¹H–¹³C HMBC Spectrum
Expansion E

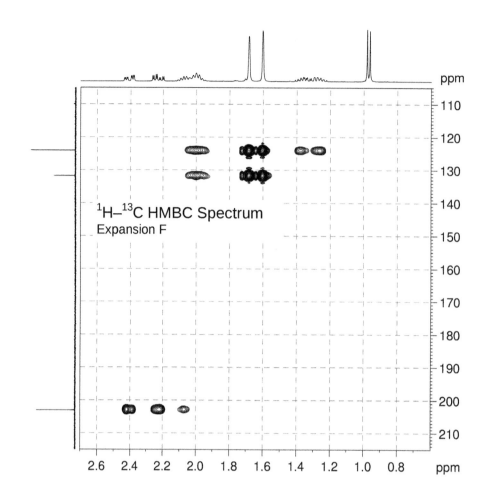

Problem 63

Identify the following compound, including any relative stereochemistry.
Molecular Formula: $C_7H_8O_2$
IR: 1765 cm^{-1}

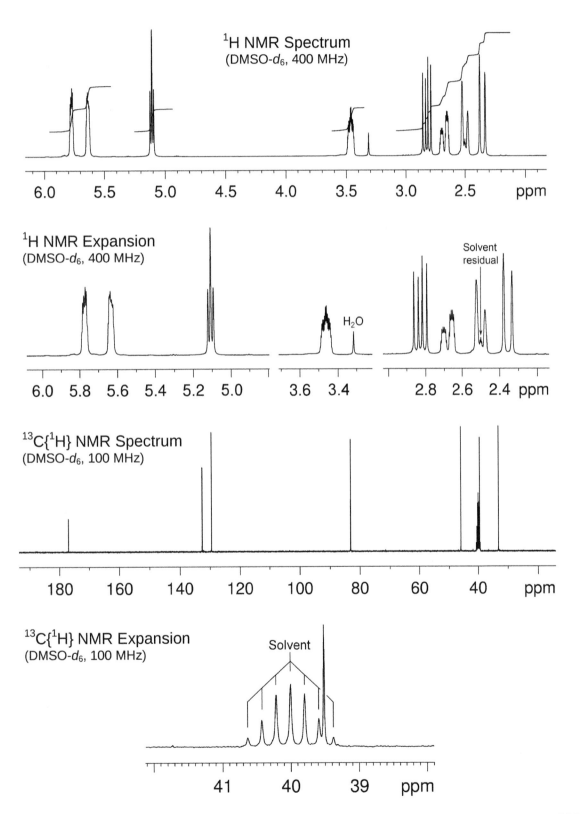

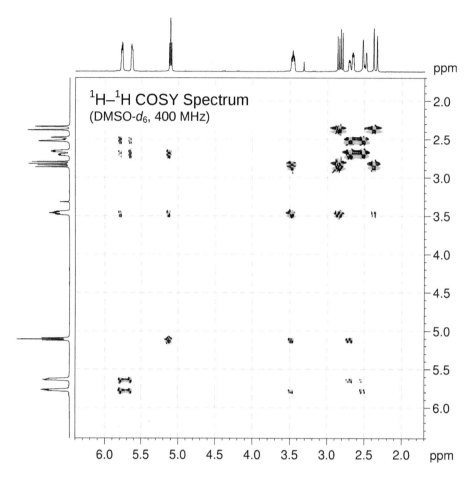

¹H–¹H COSY Spectrum
(DMSO-d_6, 400 MHz)

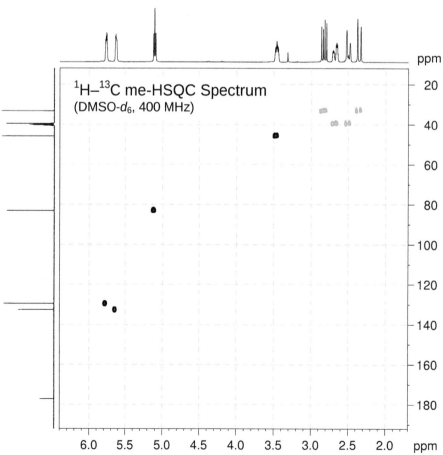

¹H–¹³C me-HSQC Spectrum
(DMSO-d_6, 400 MHz)

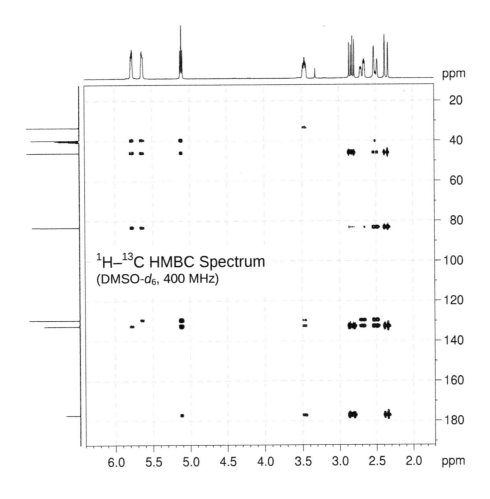

^{1}H–^{13}C HMBC Spectrum
(DMSO-d_6, 400 MHz)

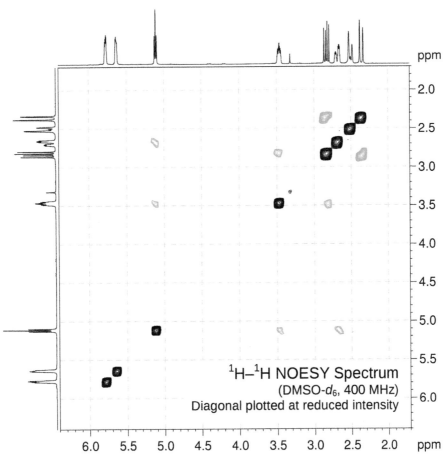

^{1}H–^{1}H NOESY Spectrum
(DMSO-d_6, 400 MHz)
Diagonal plotted at reduced intensity

Problem 64

Identify the following compound.
Molecular Formula: $C_{13}H_{16}N_2O_2$
IR: 3305, 1629, 1620, 1566 cm^{-1}

^{1}H NMR Spectrum
(DMSO-d_6, 500 MHz)

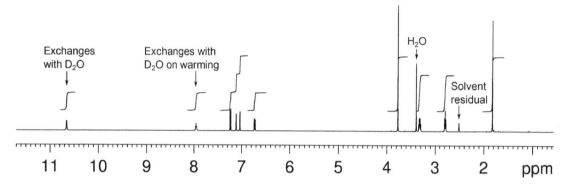

^{1}H NMR Expansion 1
(DMSO-d_6, 500 MHz)

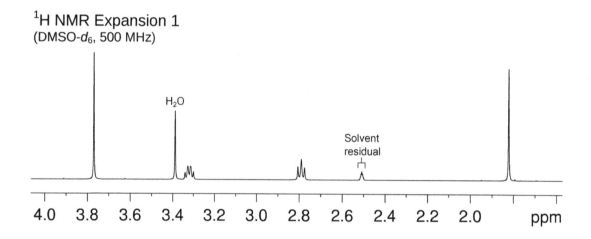

^{1}H NMR Expansion 2
(DMSO-d_6, 500 MHz)

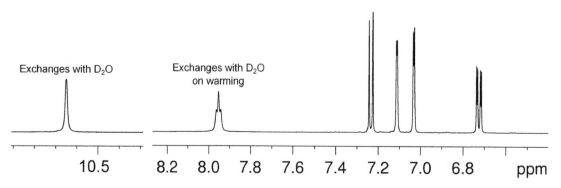

^{13}C{^{1}H} NMR Spectrum
(DMSO-d_6, 125 MHz)

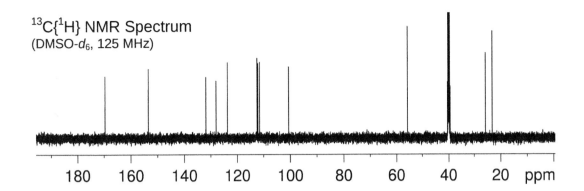

^{13}C{^{1}H} NMR Expansions
(DMSO-d_6, 125 MHz)

Solvent

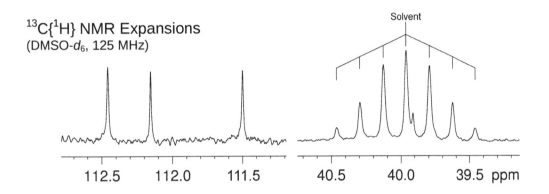

^{1}H–^{1}H COSY Spectrum
(DMSO-d_6, 500 MHz)

Expansion A

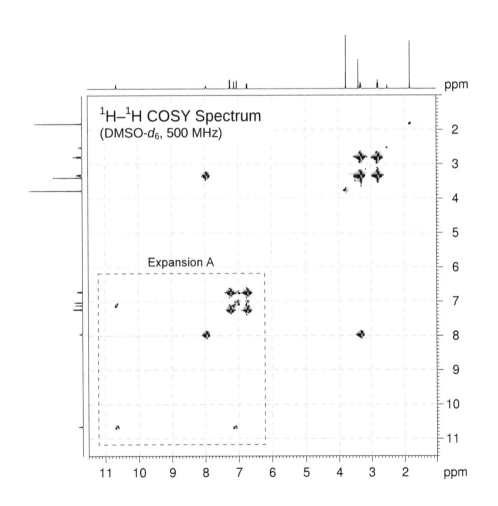

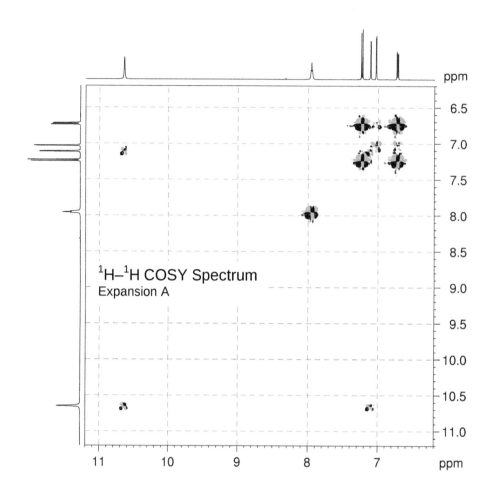

^{1}H–^{1}H COSY Spectrum
Expansion A

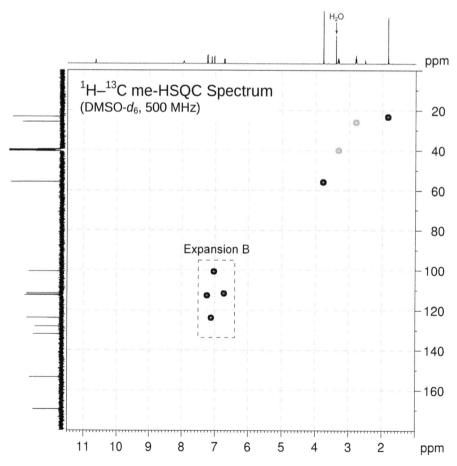

^{1}H–^{13}C me-HSQC Spectrum
(DMSO-d_6, 500 MHz)

H$_2$O

Expansion B

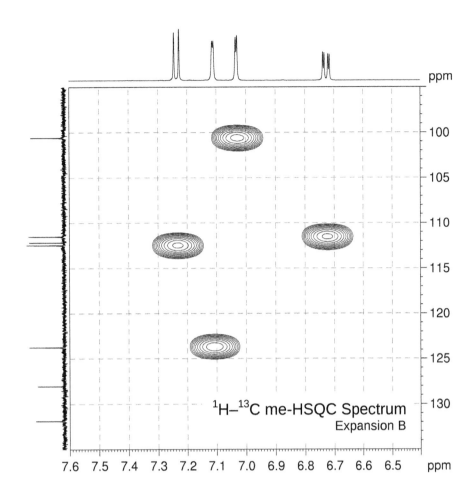

^1H–^{13}C me-HSQC Spectrum
Expansion B

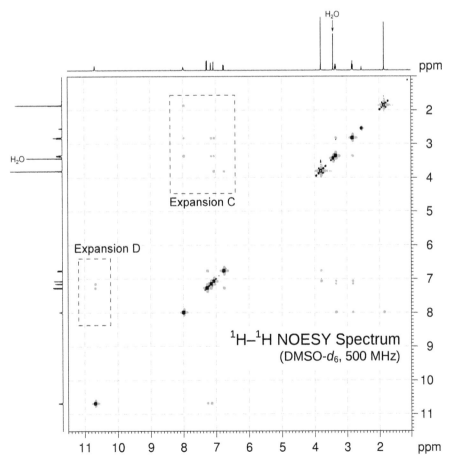

^1H–^1H NOESY Spectrum
(DMSO-d_6, 500 MHz)

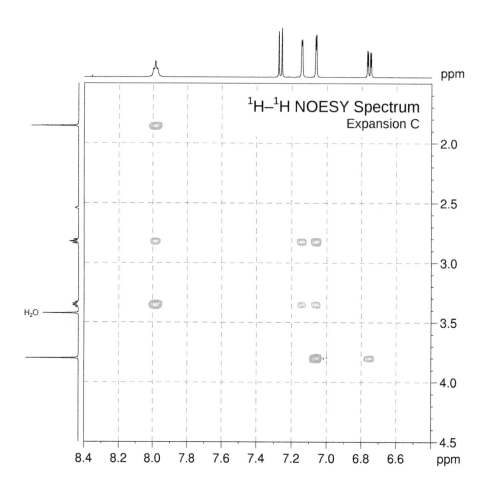

¹H–¹H NOESY Spectrum
Expansion C

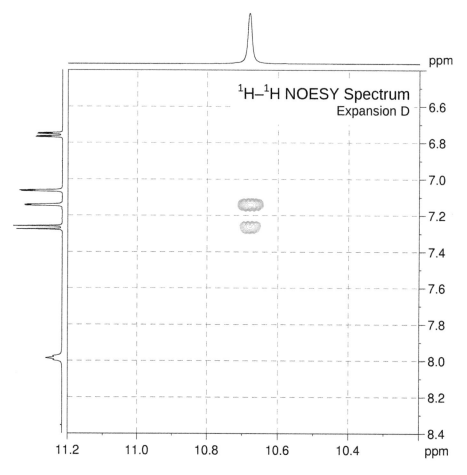

¹H–¹H NOESY Spectrum
Expansion D

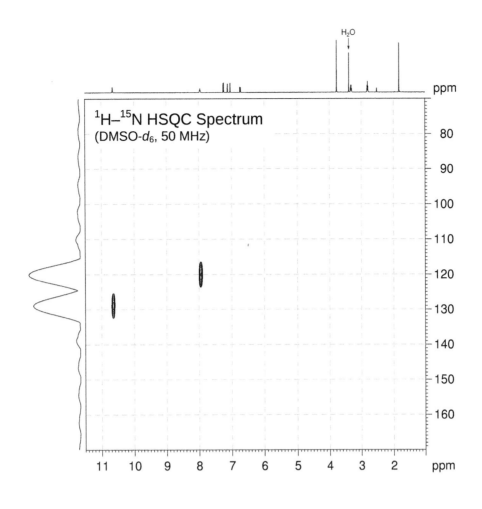

^{1}H–^{15}N HSQC Spectrum
(DMSO-d_6, 50 MHz)

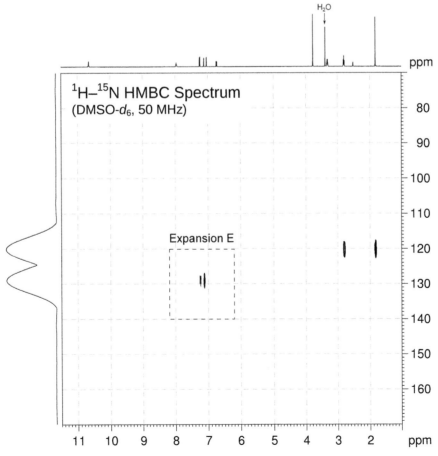

^{1}H–^{15}N HMBC Spectrum
(DMSO-d_6, 50 MHz)

Expansion E

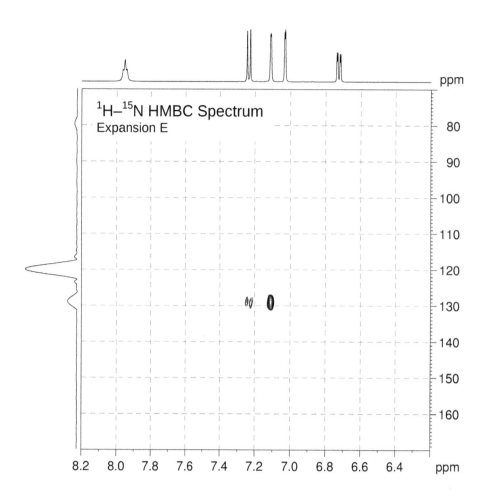

^{1}H–^{15}N HMBC Spectrum
Expansion E

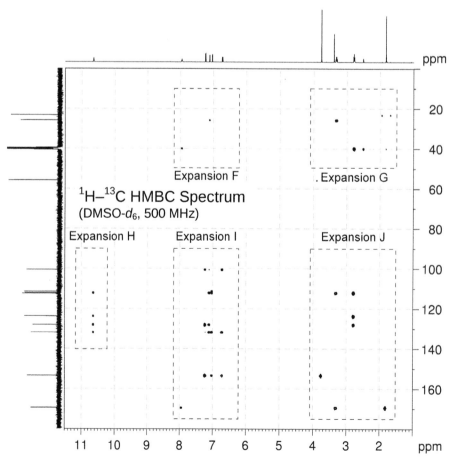

^{1}H–^{13}C HMBC Spectrum
(DMSO-d_6, 500 MHz)

Expansion F

Expansion G

Expansion H

Expansion I

Expansion J

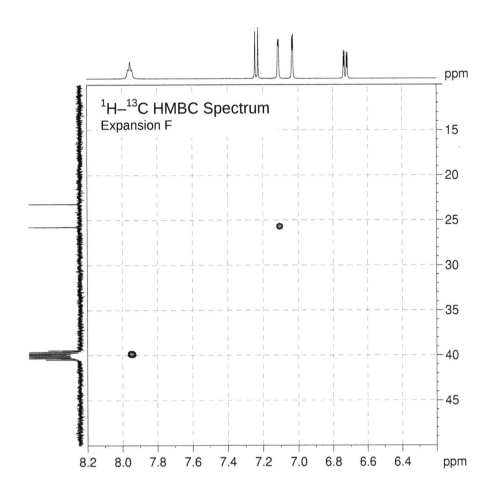

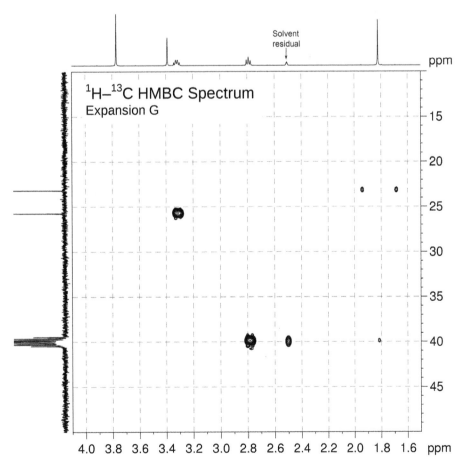

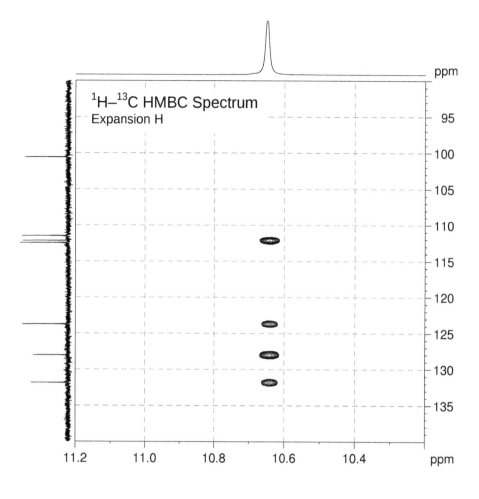

¹H–¹³C HMBC Spectrum
Expansion H

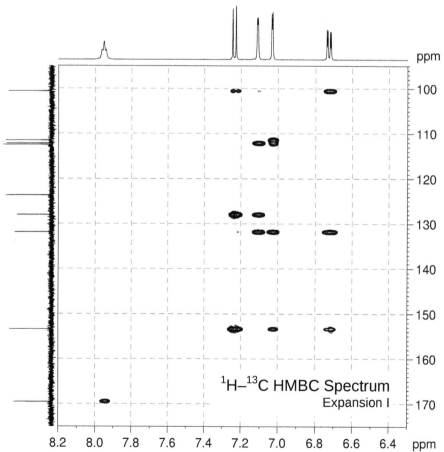

¹H–¹³C HMBC Spectrum
Expansion I

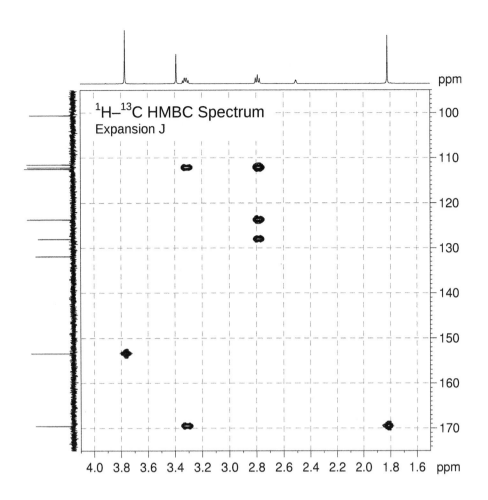

Problem 65

Identify the following compound.
Molecular Formula: $C_{10}H_{14}O$
IR: 1675 (s), 898 (s) cm^{-1}

^{1}H NMR Spectrum
(C$_6$D$_6$, 500 MHz)

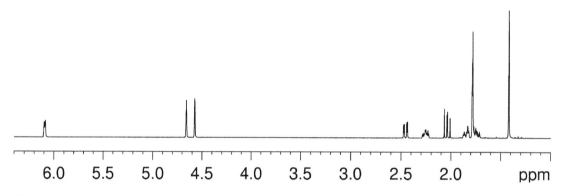

^{1}H NMR Expansion Spectrum
(C$_6$D$_6$, 500 MHz)

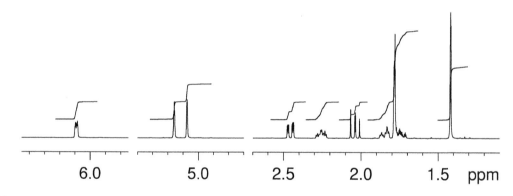

^{13}C{^{1}H} NMR Spectrum
(C$_6$D$_6$, 125 MHz)

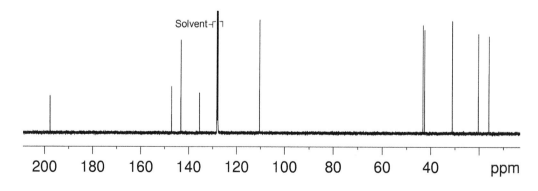

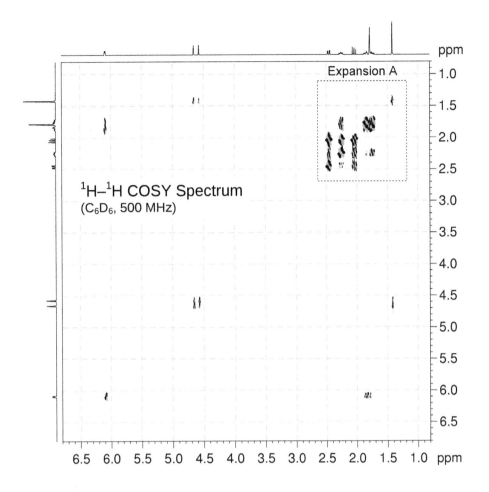

^{1}H–^{1}H COSY Spectrum
(C$_6$D$_6$, 500 MHz)

Expansion A

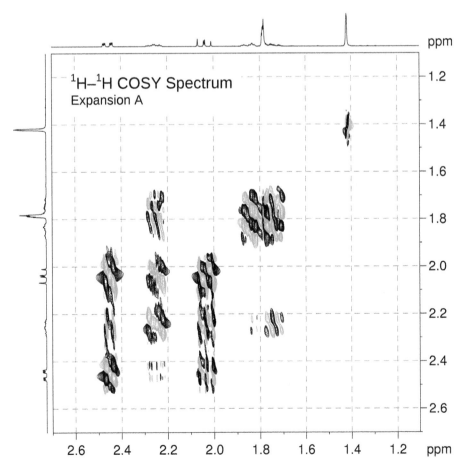

^{1}H–^{1}H COSY Spectrum
Expansion A

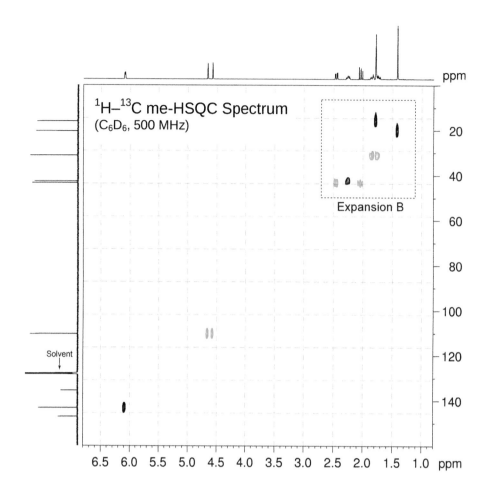

¹H–¹³C me-HSQC Spectrum
(C₆D₆, 500 MHz)

Expansion B

Solvent

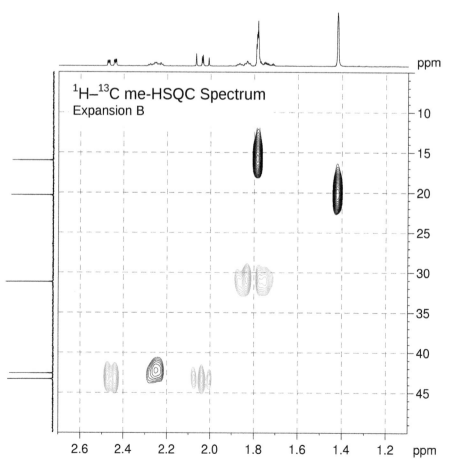

¹H–¹³C me-HSQC Spectrum
Expansion B

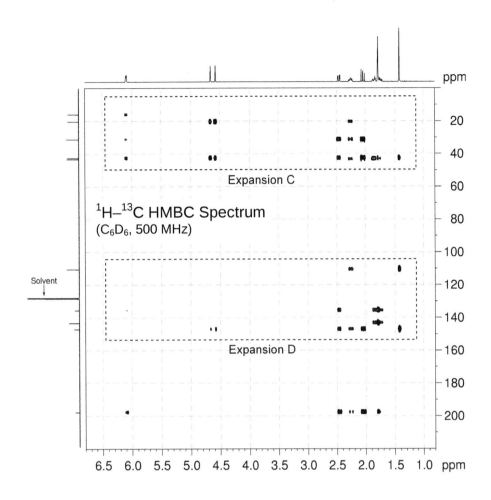

¹H–¹³C HMBC Spectrum
(C₆D₆, 500 MHz)

Expansion C

Expansion D

Solvent

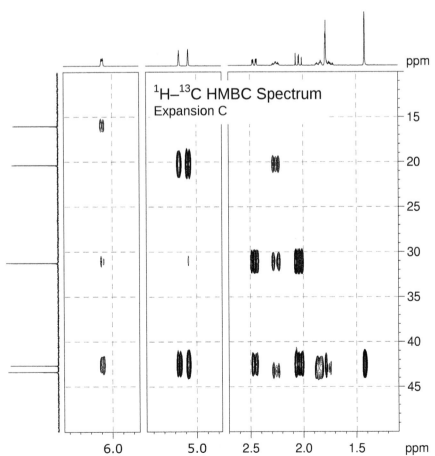

¹H–¹³C HMBC Spectrum
Expansion C

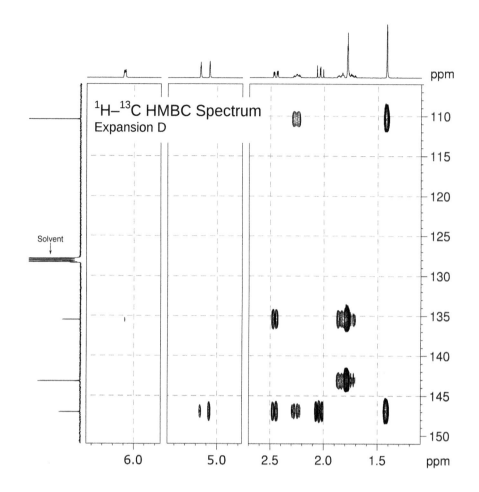

Problem 66

The ^{1}H and ^{13}C{^{1}H} NMR spectra of haloperidol (C$_{21}$H$_{23}$ClFNO$_2$) recorded in acetone-d_6 solution at 298 K and 500 MHz are given below.

The ^{1}H NMR spectrum has signals at δ 1.57, 1.81, 1.95, 2.43, 2.44, 2.67, 3.03, 3.86, 7.29, 7.31, 7.43 and 8.14 ppm.

The ^{13}C{^{1}H} NMR spectrum has signals at δ 22.3, 35.6, 38.4, 49.2, 57.6, 70.2, 115.3, 126.6, 127.8, 130.8, 131.4, 134.5, 149.1, 165.4 and 197.8 ppm.

Use the spectra below to assign each proton and carbon resonance.

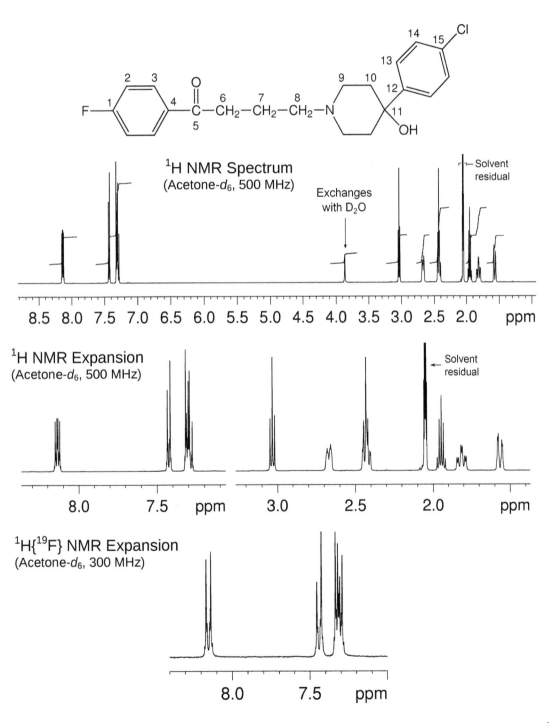

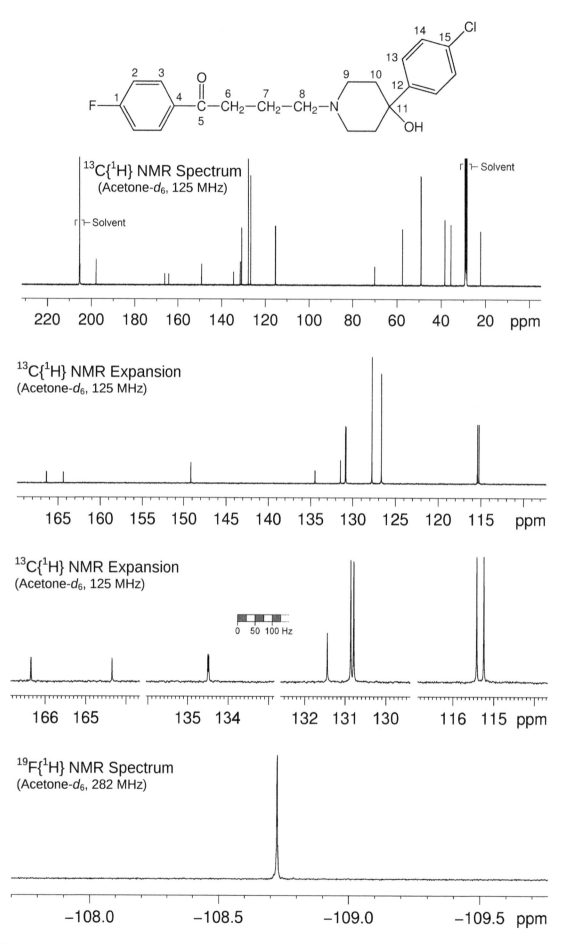

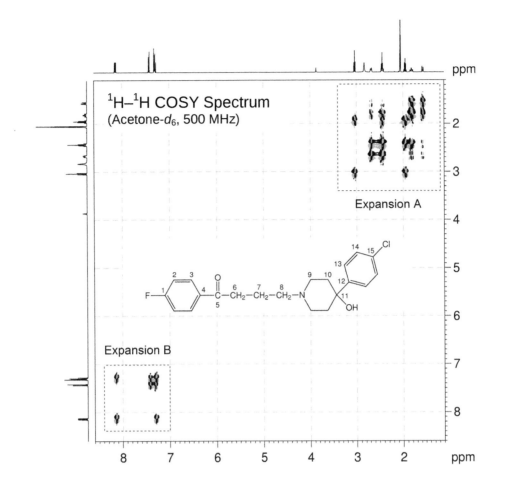

¹H–¹H COSY Spectrum
(Acetone-*d₆*, 500 MHz)

Expansion A

Expansion B

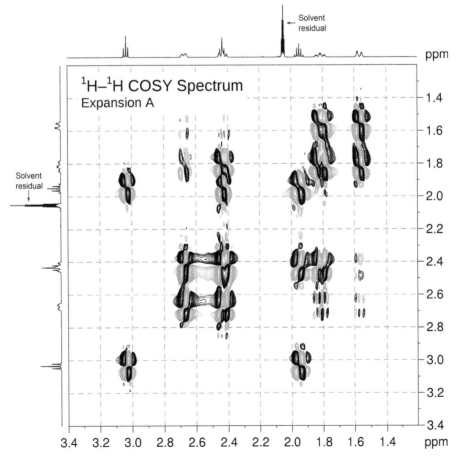

¹H–¹H COSY Spectrum
Expansion A

Solvent residual

Solvent residual

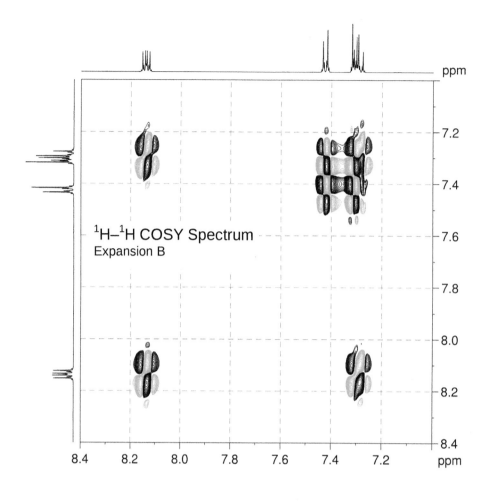

$^1H-^1H$ COSY Spectrum
Expansion B

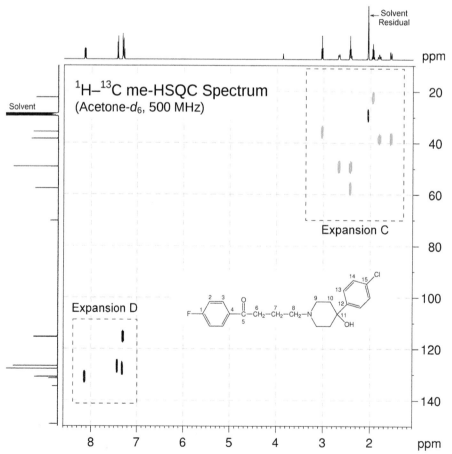

$^1H-^{13}C$ me-HSQC Spectrum
(Acetone-d_6, 500 MHz)

Solvent
Residual

Solvent

Expansion C

Expansion D

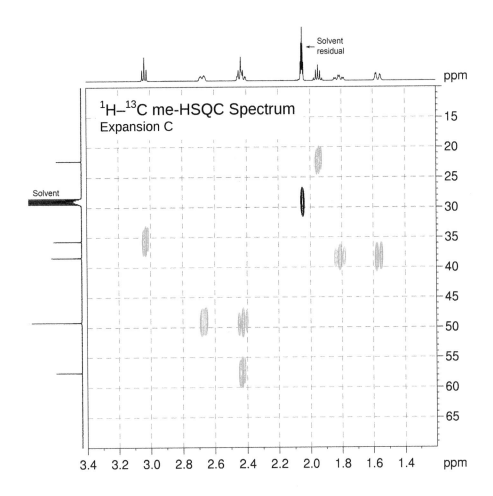

¹H–¹³C me-HSQC Spectrum
Expansion C

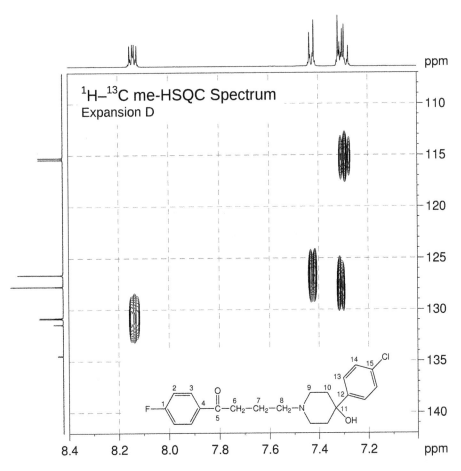

¹H–¹³C me-HSQC Spectrum
Expansion D

305

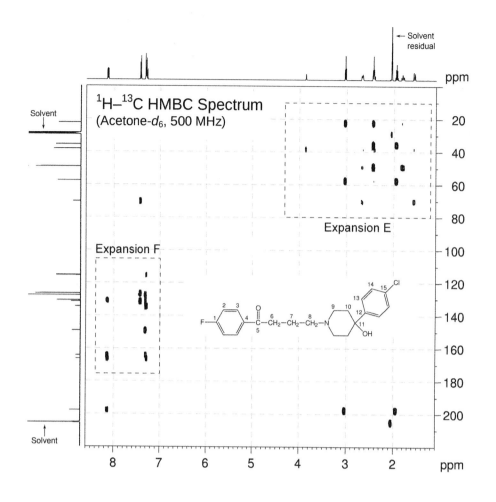

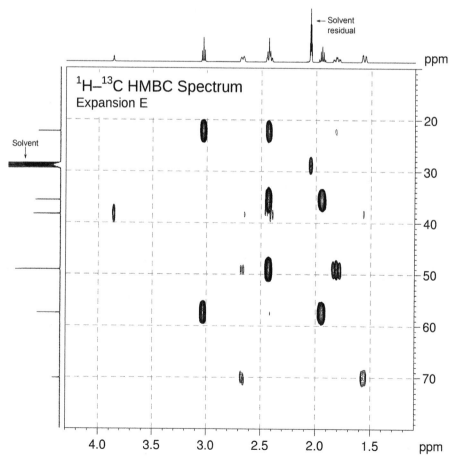

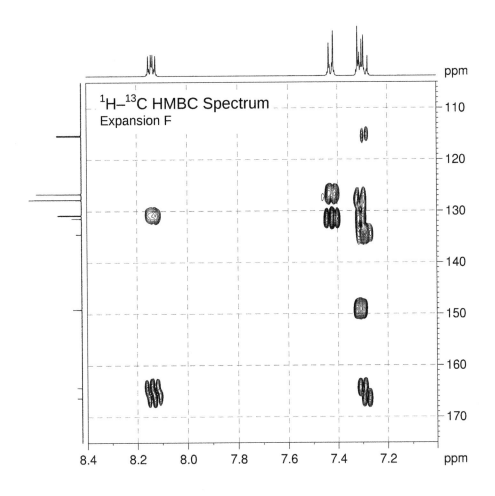

Proton	Chemical Shift (ppm)	Carbon	Chemical Shift (ppm)
		C_1	
H_2		C_2	
H_3		C_3	
		C_4	
		C_5	
H_6		C_6	
H_7		C_7	
H_8		C_8	
H_9		C_9	
H_{10}		C_{10}	
		C_{11}	
		C_{12}	
H_{13}		C_{13}	
H_{14}		C_{14}	
		C_{15}	
OH			

Index

Printed and bound by CPI Group (UK) Ltd, Croydon, CR0 4YY